Mathematics in Mind

The monographs and occasional textbooks published in this series tap directly into the kinds of themes, research findings, and general professional activities of the Fields Cognitive Science Network, which brings together mathematicians, philosophers, and cognitive scientists to explore the question of the nature of mathematics and how it is learned from various interdisciplinary angles.

The series will cover the following complementary themes and conceptualizations:

Connections between mathematical modeling and artificial intelligence research; math cognition and symbolism, annotation, and other semiotic processes; and mathematical discovery and cultural processes, including technological systems that guide the thrust of cognitive and social evolution

Mathematics, cognition, and computer science, focusing on the nature of logic and rules in artificial and mental systems

The historical context of any topic that involves how mathematical thinking emerged, focusing on archeological and philological evidence

Other thematic areas that have implications for the study of math and mind, including ideas from disciplines such as philosophy and linguistics

The question of the nature of mathematics is actually an empirical question that can best be investigated with various disciplinary tools, involving diverse types of hypotheses, testing procedures, and derived theoretical interpretations. This series aims to address questions of mathematics as a unique type of human conceptual system versus sharing neural systems with other faculties, whether it is a series-specific trait or exists in some other form in other species, what structures (if any) are shared by mathematics, language, and more.

Data and new results related to such questions are being collected and published in various peer-reviewed academic journals. Among other things, data and results have profound implications for the teaching and learning of mathematics. The objective is based on the premise that mathematics, like language, is inherently interpretive and explorative at once. In this sense, the inherent goal is a hermeneutical one, attempting to explore and understand a phenomenon—mathematics—from as many scientific and humanistic angles as possible.

More information about this series at http://www.springer.com/series/15543

Shlomo Vinner

Mathematics, Education, and Other Endangered Species

From Intuition to Inhibition

Springer

Shlomo Vinner
The Hebrew University of Jerusalem
Jerusalem, Israel

ISSN 2522-5405 ISSN 2522-5413 (electronic)
Mathematics in Mind
ISBN 978-3-319-90034-6 ISBN 978-3-319-90035-3 (eBook)
https://doi.org/10.1007/978-3-319-90035-3

Library of Congress Control Number: 2018951707

This Springer imprint is published by the registered company Springer Nature Switzerland AG
The registered company address is: Gewerbestrasse 11, 6330 Cham, Switzerland

For Hava,
my partner in life and in mathematics education

Preface

This book is a summary of my lifelong academic career in mathematics education over the course of 40 years. It is an intellectual autobiography. It represents not only my research results in the field of mathematics education but also my educational beliefs. In this book I have integrated my research in the psychology of mathematical thinking together with some attempts to teach a limited amount of middle-school and high-school mathematics, and also to explain what mathematics is. The majority of people believe mathematics is a collection of procedures that are supposed to be used in various mathematics exams at all levels. Unfortunately, the majority of adults remember mathematics as an unpleasant experience. In this book I aim to offer a corrective experience. Specifically, to tell them that if they had difficulties in studying mathematics it happened not because they were stupid but because mathematics was taught to them in a wrong way.

There are two classic books that adopted the same goal. The first one is Richard Skemp's book *The Psychology of Learning Mathematics* (1977). The second one is Robert Davis' book *Learning Mathematics: The Cognitive Science Approach to Mathematics Education* (1984). My decision was to follow Davis' model and to integrate the relevant mathematics in the relevant chapter and not in an appendix as in Skemp's book.

The book includes 17 chapters. For the convenience of the readers I am going to describe each one. Thus, they can know what to expect and in case they are not interested in a given chapter they can skip it.

I hope this book suggests a corrective experience to some readers who have negative memories from their high-school mathematics. However, if they are not interested in such an experience they are advised to skip these chapters and to focus on the general-education chapters.

Chapter 1. Is Mathematics Required in Real-life Situations?

In this chapter I argue against the common claim that mathematics is required in real-life situations.

Chapter 2. What Is Mathematics for?

In this chapter I discuss the rationale of teaching mathematics to the entire school population. The intention is to develop the mind of the student. However, the way it is commonly taught fails to achieve that. The way it is commonly taught prepares the students to pass various exams. Moreover, mathematics serves as a filter for further studies. Thus students who are weak in mathematics are left behind even if they are brilliant in other domains like literature, history, and art.

Chapter 3. Can Scientific Thinking Help Us Cope with Our Existential Problems? Can Religious Thinking Help Us Cope with Our Existential Problems?

In this chapter I discuss the contribution of mathematical thinking, as part of scientific thinking, to coping with our existential problems. I also discuss the contribution of religious thinking to coping with our existential problems. In my opinion, our main existential problems are mortality and suffering (both psychological and physical). Thus, the nature of this chapter is philosophical-psychological.

Chapter 4. Concept Formation in Mathematics: Concept Definition and Concept Image

In this chapter I discuss concept formation in mathematics. On one hand, mathematical concepts (in case they are not primary concepts) should be formed by their mathematical definitions. However, the practice shows that, generally, with the majority of people, this is not true; concepts are formed by means of their examples.

Chapter 5. The Misconception Fallacy, the Pseudo-conceptual and the Pseudo-analytical Behaviors in Mathematical Contexts

In this chapter I examine the commonly accepted claim that each mistake people make is a result of a misconception. My claim is that there are other modes of thinking that form mistakes. The names I give them are the *pseudo-conceptual* and the *pseudo-analytical* modes of thinking. People who are involved with these modes of thinking in certain situations pick up some notions that look relevant to the context in question, and use these notions in their responses in a meaningless way.

Chapter 6. The Pseudo-conceptual Behavior in Contexts Outside the Mathematics Classroom

In this chapter I show that the *pseudo-conceptual* mode of thinking also occurs in situations in philosophy and social sciences. Thus, we see that it is wrong to restrict certain modes of thinking to mathematical situations. Mathematical thinking is not an "isolated island."

Chapter 7. Can E-learning (Computer-based Learning) Replace Teacher–Student Individual Communication?

My answer to the above question is negative. In order to establish my claim I am relying on a well-known work by Erlwanger (1973), which shows that without intensive interaction with a teacher the pupil can reach huge misconceptions about the topic under consideration.

Chapter 8. What Is Mathematics?

This chapter has a reflective-philosophical nature. Usually people do not ask themselves what mathematics is. However, they have implicit ideas about it. The most common idea, in my opinion, is that mathematics is a collection of procedures (algorithms, formulas, etc.) intended to be used in solving various examinations during their mathematics studies.

I explain that mathematics is a collection of theories (number theory, group theory, game theory, for example). Each theory has the following structure: a set of abstract objects, relations on this set, as well as operations on the set. Each theory has a deductive system, including axioms and rules for proving theorems. To clarify this complicated definition, I use the example of the theory of school arithmetic. Most of this structure is already well known to pupils in the sixth grade. I describe its axioms, including the axiom of mathematical induction, and use the axioms to prove some of its theorems.

Chapter 9. The Role of Examples in the Learning of Mathematics and in Everyday Thought Processes

In this chapter I discuss the role of examples in the learning of mathematics and in everyday thought processes. I discuss concept formation, conjecture formation, and conjecture verification in mathematics and in everyday thought processes. Thus, I emphasize again that the domain of mathematical thinking is not an "isolated island." It is strongly connected to the domain of everyday thought processes.

Chapter 10. Imagination Acts

In this chapter I discuss what I suggest calling "imagination acts." These acts are strongly needed in mathematical thinking, which requires us to imagine things that do not exist in our real life. For instance: In Euclidean geometry we are required to think of points that have no dimensions, about infinite lines, and so on. These things do not exist in the real world. In the real world, points have a certain thickness and lines are finite. Thus, people who are not capable of these imagination acts might have difficulties in studying some mathematical topics.

Chapter 11. The Wonders of Infinity and Its Weirdness

The reason why this chapter follows the chapter about imagination acts is quite obvious. In order to follow it, the readers have to imagine a hotel whose rooms are located along an infinite half line. They have to imagine an infinite number of guests coming to stay in this weird hotel.

The chapter starts with questions like: Which set is bigger, the set of all natural numbers or the set of prime numbers? On one hand, it is obvious that the set of all natural numbers is bigger because it contains the set of prime numbers. I show that this is not the case.

Moreover, it is possible to show that the set of all rational numbers has also, in a way, the same size as the set of all natural numbers. Thus, one can get the impression that the set of all real numbers (the set of numbers that high-school students deal with, which includes the rational as well as the irrational numbers) has the same size as the set of all natural numbers. However, this impression is wrong. Showing that requires serious mathematical effort, both notational and conceptual. I believe that readers who studied mathematics in an accelerated track in high school are capable of following it. However, readers who do not want to undertake this mathematical challenge are advised to skip this chapter.

Chapter 12. Mathematics, Educational Values, and Intellectual Values

In this chapter I emphasize my conception that the mathematics teacher should also be an educator and not simply a person who covers the course syllabus. I discuss the moral value people call "the golden rule." I also discuss the educational value of procedures. It is important to follow mathematical procedures (like order of operations, algorithms, and so on). However, it is also important to follow procedures in the real world. Some of the procedures are dictated by law. For instance: not driving through a red light. Other procedures are only recommendations. For instance: turning off lights and air conditioning when leaving home. If you do not follow this recommendation the result is financial (not a big deal). It is not disastrous as it would be in case you drove through a red light. At the end of the chapter I point out some intellectual values that can be learned in the mathematics classroom.

Chapter 13. What Are Elementary Mathematics Teachers Expected to Know?

In this chapter I discuss the possible answers to the above question. First of all I recommend to them not to be arrogant (unfortunately, it still happens), and not to disparage pupils who have difficulties in mathematics. These students can be brilliant in other domains. Moreover, it is important to make teachers aware of the reasons for teaching mathematics.

Of course, teachers are supposed to master the mathematics they teach (content knowledge). They are also expected to know pedagogical knowledge. They are expected to understand the typical mistakes of their students. I try to draw the profile of a typical elementary mathematics teacher based on many teachers that I have met as a teacher trainer. I found them dedicated people who love kids and love to help their intellectual and emotional development.

It is unfair to them to criticize their weakness in mathematics. One should consider their contribution to the emotional and the intellectual development of their pupils. If you consider that, then their weakness in mathematics is negligible.

Chapter 14. Proofs by Contradiction or Isn't This an Absurdity?

The nature of this chapter is mathematical again. Usually, at least some students face proofs by contradictions in their high-school courses. For many of them it is a problem. I try to explain the reason for these difficulties. I make an effort to simplify it by presenting some proofs by contradiction. One of them is the claim that $\sqrt{2}$ is irrational. The second one is the claim that there are infinitely many prime numbers.

I point out difficulties some people may have when asked to assume a counter-reality situation. I chose to demonstrate it by means of a hilarious piece of literature: *The Lesson* by Ionesco (1951).

Chapter 15. Visually-moderated Sequences

I first met this concept in Robert Davis' book (1984). It relates to procedures (sequences of steps). While performing a certain step you do not foresee all steps of the procedure. However, each step shows you the way to the following step. It happens many times in mathematics (I point out some examples) but it also happens a lot outside mathematics. For instance, when driving we do not remember all the turning points on our way. However, whenever we reach a turning point, we visually recognize it, and we remember to make the turn. I point at some examples where visually-moderated sequences go wrong.

Chapter 16. Science Versus Common Sense and Theology

In this chapter I point out the fact that sometimes there is a conflict between science and common sense (in Fischbein's words, primary intuitions). The first conflict is in the impression that the sun moves during the day from east to west. It took several centuries until Copernicus came up with his theory that the sun is stationary and our planet Earth is rotating.

His theory was rejected by the Church. According to the Church our planet was the center of the world. Copernicus' theory was a blow to this belief. He was not persecuted because his book was published after his death. However, one of the greatest scientists of all ages, Galileo Galilei (1564–1642), was a victim of the Church persecution. He was not allowed to teach the theory of Copernicus, and later on he had to spend his last years under house arrest. Eventually, Copernicus' theory was accepted. We learned to be humble. This was not so easy. It is hard to accept that you are not the center of the world. Therefore, it took some years before the Church agreed to accept Copernicus' theory.

Several centuries later, in 1925, a high-school teacher was taken to court because he taught Darwin's theory of evolution. He was found guilty and had to pay a $100 fine. Fortunately, the inquisition did not exist in 1925. Humanity became more tolerant.

Chapter 17. Rationality

I chose to conclude my book with the notion of rationality. This is because I consider mathematics and science products of rational thinking. Thus, to behave rationally implies to behave according to the recommendations of science. (For example, to avoid smoking. It took some centuries to find out that smoking is dangerous to our health. Other examples are to avoid junk food and to do physical exercises.) One should remember that science keeps changing through time. Thus, being rational in the twenty-first century is different from being rational in the days of Newton.

But rationality is more than that. Here is an excerpt from Merriam-Webster's characterization of *rational*: "To be rational is to be reasonable. Rationality is the quality or state of being agreeable to reason. Rationality is applied to opinions, beliefs and practices." About being reasonable, the dictionary adds that reasonable is "not extreme or excessive" and it is "moderate and fair."

Wikipedia's description of *rational* includes: "generally positive attributes, including: reasonable, not foolish, sane and good."

Both Merriam-Webster and Wikipedia agree that rationality, in ordinary language, also has a moral aspect (moderate, fair, and good). Be fair! Be considerate! Be a human being!

About the Subtitle "From Intuition to Inhibition"

The phrase "from intuition to inhibition" expresses one of the main features of mathematical thinking. As human beings it is quite natural for us to be spontaneous or even impulsive. That is the intuitive mode of action. Very often it is quite a good start. However, mathematical thinking recommends examining our intuitions. Sometimes they are correct. Sometimes they are wrong; therefore it is recommended to inhibit them. My suggestion is that this recommendation be applied also to everyday life. Explicitly, in everyday life it is also useful to control spontaneous and impulsive reactions.

The Potential Readers of the Book

Although this book requires knowledge of a limited amount of algebra in order to understand some parts, it is also about education. Hence, readers who are not interested in mathematics but are interested in education are invited to read the parts that do not require mathematical knowledge.

An Opportunity for a Corrective Experience for Those Who Carry Negative Memories from Their Mathematics Learning in School

As I said earlier, this book offers an opportunity for a corrective experience for interested readers. Learning mathematics in all school levels was, unfortunately, an unpleasant experience for the majority of people. This can be understood if we look at the way mathematics is taught by many teachers. Some of them are arrogant, thinking that if you do not understand immediately you are stupid. This is especially true of the attitude some mathematics teachers have towards girls. One Israeli teacher said to one of his female students: "If a bird had your brain, it would fly backwards." This girl became one of the leading Israeli poets. Another teacher said to one of his female students after she gave him a wrong answer: "Well, for changing diapers and cooking you have the required talent." Some teachers who belonged to the previous generation lacked the pedagogical tools to explain mathematics to their students. They were not aware of the students' difficulties. I believe and hope that at least 50 years of research in mathematics education have changed the situation. Yet, these 50 years of research are not adequately exposed to students who prepare themselves to teach mathematics. Usually, they are exposed to them in master's degree programs in mathematics education.

Hence, I did my best to teach interested readers some mathematical topics using the accumulated pedagogical knowledge about concept formation. I believe that these topics are fascinating, but they are usually not part of the school curriculum. However, they can easily be understood by tenth graders. One such topic is infinity.

About Reflection

Undergoing the above-mentioned track will give readers an opportunity to reflect on their past mathematical experience and on their mathematical thought processes. For me, reflection is a positive thought process. The reflective person is more moderate, more rational, and more considerate than one who is not reflective. Thus, I can name my recommendation for reflection "a call for reflection."

The Hebrew University in Jerusalem, Israel Shlomo Vinner

Acknowledgments

I would like to thank all the people without whose support this book would not have been written. First of all, Marcel Danesi who invited me to write this book. The moment I gladly accepted his invitation, Elizabeth Loew took over and accompanied all the stages of the writing of the book. The book reviewer, Reuben Hersh, made many useful suggestions that improved the book immensely. I am deeply grateful to him for his serious reading of the entire book from its first letter to its last one.

I would like to thank Dahlia Fisch for her careful language editing of the book. She taught me a great deal of English grammar which I was not aware of since I studied English with my high school teacher. Particularly, I learned from her about the difference between "which" and "that."

And last, not least, special thanks to Hava, my wife, who carefully read the book and pointed out to me some places that were clear to me when I wrote them, but were not clear to her when she read them. Thus, she helped me reformulate them to be clear and simple. Therefore, practically, she became an editor of the book.

Moreover, the hardest task in the book production was the index. For this task Hava recruited all her intellectual abilities and her computer skills. Without her contribution I cannot see how the book would have an index.

Finally, I would like to thank my many students who were part of my mathematics education research, who led me to the insights I have about the learning and the teaching of mathematics.

Contents

Chapter 1
Is Mathematics Required in Real-life Situations?

One opinion of mine (which is quite controversial, I must admit) is about the importance of mathematics in real-life situations. I claim that the use of mathematics in common everyday-life situations is quite limited. We need mathematics in situations of counting, giving change, and calculating tips in restaurants. Sometimes, in order to support my claims about the necessity of mathematics in everyday-life situations I prefer to refer to great writers or composers. Regarding the algebra people need, I refer my audience to Mozart's Catalogue Aria from *Don Giovanni*. I used to show a clip of this Aria in my lectures, something I cannot do in my book. Here I can only present its text. (Of course, it is more enjoyable to listen to it and to see it when sung by Don Giovanni's servant Leporello. I recommend to my readers to watch it on YouTube or elsewhere.)

Here is the text of the Aria:

> My dear lady, this is the list
> Of the beauties my master has loved,
> A list which I have compiled.
> Observe, read along with me.
>
> In Italy, six hundred and forty;
> In Germany, two hundred and thirty one;
> A hundred in France; in Turkey, ninety-one;
> But in Spain already one thousand and three.
>
> Among these are peasant girls,
> Maidservants, city girls,
> Countesses, baronesses,
> Marchionesses, princesses,
> Women of every rank,
> Every shape, every age.
>
> With blondes it is his habit
> To praise their kindness;

S. Vinner, *Mathematics, Education, and Other Endangered Species*, Mathematics in Mind, https://doi.org/10.1007/978-3-319-90035-3_1

In brunettes, their faithfulness;
In the fair-skinned, their sweetness.

In winter he likes fat ones.
In summer he likes thin ones.
He calls the tall ones majestic.
The little ones are always charming.

He seduces the old ones
For the pleasure of adding to the list.
His greatest favorite
Is the young beginner.

It doesn't matter if she's rich,
Ugly or beautiful;
If she wears a skirt,
You know what he does.

By the way, Don Giovanni's behavior is a classic example for immoral behavior. This was also Da Ponte's point of view, the author of Don Giovanni's libretto. This was expressed by the subtitle of the opera: "The Libertine Punished."

As to the need of geometry in everyday life, in my lectures I used to bring up the opening aria of *The Marriage of Figaro* by Mozart. Here, we watch Figaro measuring the size of the bed that he and his future wife, Suzanna, are going to get for their marriage. Figaro is interested to find out whether the size of the bed fits the size of his future wife. He counts: "five, ten, twenty." Suzanna, on the other hand, is interested in her wedding dress. Here, Da Ponte (who was also the librettist of The Marriage of Figaro) points at the difference between men and women at the evening of the marriage. The man is interested in the bed. The woman is interested in her dress.

The above examples expressed my opinion about the need for mathematics in everyday life. My claim is that we need a very limited amount of it in everyday life. This claim has been, surprisingly, supported by a research mathematician, Underwood Dudley, who has asked: "What is mathematics for?"

In fact, his question was: "What is mathematics education for?" (Dudley, 2010). His claim was that it is not in order to use it in everyday life. Its main goal is to train the mind. For him, training the mind means to be able to apply what you have learned to all kinds of new questions.

His main claim is that mathematics is to teach reasoning, usually through the medium of "silly problems." Dudley presents in his paper some of these "silly problems." Here is one of them:

> In the Rhind Papyrus, that Egyptian textbook of mathematics c. 1650 BC, we find: Give 100 loaves to five men so that the shares are in arithmetic progression and the sum of the two smallest is 1/7 of the three greatest.

Here, I would like to remind the reader what an arithmetic progression is. It is a sequence in which every element excluding the first one is obtained from its predecessor by adding a constant. If we denote the elements of the sequence by a_1, a_2, a_3, …, a_n and the constant by d, then the nth element of the arithmetic progression is: $a_n = a_1 + (n - 1)d$. It is quite clear that the ancient Egyptians did not use the term "arithmetic progression." Dudley uses it in order to clarify and simplify the question to the contemporary reader. Translating the above question to the language of algebra, we get: $a_1 + a_2 = \frac{1}{7}(a_3 + a_4 + a_5)$.

If we use the above formula for the nth element of an arithmetic progression, we get the equation:

$$a_1 + (a_1 + d) = \frac{1}{7}((a_1 + 2d) + (a_1 + 3d) + (a_1 + 4d)).$$

After removing the parentheses and combining like terms we obtain:

$$2a_1 + d = \frac{1}{7}(3a_1 + 9d).$$

Multiplying the two sides of the equation by 7, we get:

$$14a_1 + 7d = 3a_1 + 9d; 11a_1 = 2d; d = \frac{11}{2}a_1.$$

In addition, we have: $a_1 + a_2 + a_3 + a_4 + a_5 = 100$.

After substituting a_1, a_2, a_3, a_4, a_5 by means of the arithmetic progression formula, we get: $a_1 + (a_1 + d) + (a_1 + 2d) + (a_1 + 3d) + (a_1 + 4d) = 100$, namely, $5a_1 + 10d = 100$; dividing it by 5, we get: $a_1 + 2d = 20$.

Earlier we obtained: $d = \frac{11}{2}a_1$.

After substituting it in the above equation, we get:

$$a_1 + 11a_1 = 20; 12a_1 = 20; a_1 = \frac{20}{12} = \frac{5}{3}; d = \frac{11}{2}a_1 = \frac{11}{2} \cdot \frac{5}{3} = \frac{55}{6}.$$

Therefore, the shares are: $1\frac{2}{3}, 10\frac{5}{6}, 20, 29\frac{1}{6}$, and $38\frac{1}{3}$.

Note that I have tried not to skip any step while solving the above equations. An experienced mathematician might be quite bothered by this. However, my purpose was to show that even people who studied a minimal amount of high-school algebra will be able to follow the above procedure, even many years later.

Another comment: We have no idea how the ancient Egyptians obtained the above result. They could not have used algebra, which was invented in the sixteenth century. However, they had quite sophisticated mathematics. Dudley goes on with his comments: "The ancient Egyptians were a practical people, but even so, this eminently unpractical problem was thought to be worth solving."

At this stage, I would like to pose an intermediate question to my readers: "What is more important to our adult life, to remember the formula for the nth element of an arithmetic progression or not to smoke?" As a teacher, I always recommended to my master's students to stop smoking. When a student came to me at the end of the course and said to me, "I want to tell you that I have stopped smoking," I considered it one of my major educational achievements. On the other hand, there are still teachers and parents who smoke. They are negative role models for their students and children.

After the intermediate comment, I would like to get back to Dudley's paper in order to convince those who are still not convinced about the role of mathematics in everyday-life situations.

Dudley presents some textbook problems and discusses the question of whether they are really needed to help people in real-life situations:

1. "You are a facilities manager for a small town. The town contains approximately 400 miles of road that must be plowed following a significant snowfall. How many plows must be used in order to complete the job in one day if the plows can travel at approximately 7 miles per hour when engaged?"
2. "How much ice cream mix and vanilla flavor will it take to make 1000 gallons of vanilla ice cream at 90% overrun with the vanilla flavor usage rate at 1 oz. per 10 gallon mix? (90% overrun means that enough air is put into the frozen mix to increase its volume by 90%.)"

   And Dudley adds: "Though dressed up with x's and y's, the solution amounts to calculating that you need $1000/(1 + 0.9) = 526.3$ gallons of mix to puff up into 1000 gallons of ice cream, so you will need $526.3/10 = 52.6$ ounces of flavor. The employee adding the flavor will not need algebra, nor will he or she need to think through this calculation. There will be a formula, or a rule that gives the result, and that is what happens on the job. Problems that arise on the job will be for the most part problems that have been solved before, so new solutions by workers will not be needed."
3. "An investment club decided to buy $9000 worth of stock with each member paying an equal share. But two members left the club, and the remaining members had to pay $50 more apiece. How many members are in the club?"

   And Dudley asks: "Is this really a real-life situation?"

As for using geometry and trigonometry that Dudley has learned in high school, he makes the following comment:

> This was maybe the second time in my life—maybe the first—that I had occasion to use the geometry and trigonometry that I had learned in high school. Furthermore, I had never had occasion to use the higher mathematics that the high-school math had prepared me for. Never! Why did I—and millions of other students—spend valuable educational hours learning something that we would never use? Is this education? Learning skills that we will never need?
>
> The purpose of learning math, which most of us will never use, is only to prepare us for further math courses—which we will use even less frequently than never.

Dudley quotes J. D. Fitch who wrote the following in 1906:

> Our future lawyers, clergy, and statesmen are expected at the University to learn a good deal about curves, and angles, and number and proportions; not because these subjects have the smallest relation to the needs of their lives, but because in the very act of learning them they are likely to acquire that habit of steadfast and accurate thinking, which is indispensible in all the pursuits of life.

Dudley's final remark is:

I am not so unrealistic as to advocate that textbook writers start to produce texts with titles like algebra, a Prelude to Reason. That would not fly. We do not want to make unwilling students even more unwilling. . . .Can we be a little less insistent that mathematics is essential for earning a living? What mathematics education is for is not for jobs. It is to teach the race to reason. It does not, heaven knows, always succeed, but it is the best method that we have.

It is not the only road to the goal, but there is none better. Furthermore, it is worth teaching.

Were I given to hyperbole I would say that mathematics is the most glorious creation of the human intellect, but I am not given to hyperbole so I will not say that. However, when I am before a bar of judgment, heavenly or otherwise, and asked to justify my life, I will draw myself up proudly and say, 'I was one of the stewards of mathematics, and it came to no harm in my care.' I will not say, 'I helped people get jobs.'

Chapter 2
What Is Mathematics for?

I have borrowed the title of the above Dudley's paper for this chapter in order to tell my readers that besides the task assigned to mathematics education by Dudley, in educational systems mathematics has an additional task: To be a filter for further studies. The first mathematics educator who expressed it clearly and bravely was (to the best of my knowledge) Jere Confrey (1995). There (p. 3) she claimed: "In the vast majority of countries around the world, mathematics acts as a draconian filter to the pursuit of further technical and quantitative studies." Confrey uses the adjective "draconian", which has a negative connotation. I would like to avoid this negative connotation and in a moment I will explain why. Unfortunately and fortunately, not everybody has the intellectual faculties to study at a college. This sounds cruel and even arrogant, but (if I am allowed not to be politically correct for a moment), our society needs, in addition to college graduates, taxi drivers, cashiers, waiters and waitresses, building constructors, janitors, garbage collectors, house cleaners, street cleaners, security guards, and many other simple workers. These workers will not be college graduates. Only in rare cases, college graduates will prefer jobs which do not require college education. This may happen for several reasons. One of them is that the salaries for certain jobs are lower than the salaries for other jobs that require college studies. For instance, in several countries, including my country, the salary of a teacher is lower than the salary of a bank clerk. Thus, although teaching is a more interesting profession (at least in my opinion) than being a bank clerk some people will prefer to be a bank clerk. They will say that they need money to support their families and they are right. However, if the teaching salary is enough for the basic needs it is better, at least for me, to be a teacher than to be a bank clerk, which looks to me quite a boring job. The teaching profession is one of the most interesting professions because it gives the teacher a real and deep communication with people. The bank clerk has only superficial communication with people. However, some people do not like to have a real and deep communication with other people at work. And this might be the reason some people prefer jobs like bank clerk, security guard, etc.

S. Vinner, *Mathematics, Education, and Other Endangered Species*,
Mathematics in Mind, https://doi.org/10.1007/978-3-319-90035-3_2

Let us get to the question whether mathematics is a fair filter. Well, being fair is relative. Do we have better filters? I believe we don't. Let us assume that we suggest writing a composition as a filter. Evaluating compositions is too subjective. One of our most famous writers was asked by his son to write his composition for him. The father, as a joke, undertook the challenge. The literature teacher gave him a C. On the other hand, the grade on a mathematics test is quite objective. The examiners are clearly instructed how to deduct points. This many points for calculation mistakes, this many points for conceptual mistakes, for logical mistakes, and so on. However, using mathematics for students who are interested in studying topics in which mathematics is not required (history, literature, art, music, etc.) is quite unfair.

The optimal solution (I am not claiming that it is a good solution) is to minimize the mathematical requirements for these students. In other words, to require only the minimal mathematics curriculum needed in order to get a matriculation certificate or (in the USA) to pass the scholastic aptitude test (SAT).

Chapter 3
Can Scientific Thinking Help Us Cope with Our Existential Problems? Can Religious Thinking Help Us Cope with Our Existential Problems?

I have decided to present the chapter that discusses existential problems now, and present my work in mathematics education as a researcher and teacher in subsequent chapters.

In my lifelong career as a science educator, I have tried to convince my students to adopt the scientific way of thinking. Hence, I have asked myself to what extent the scientific way of thinking helps people cope with their main existential problems. I was also led to ask myself what the main existential problems are. I am not sure that my answers will be acceptable to everybody. I admit that what I present here is quite subjective. However, I believe that when discussing emotions of human beings philosophically, we cannot avoid subjectivity.

When dealing with thinking modes in the context of existential problems, one cannot ignore religious thinking, which is quite common in all societies. Thus, I have also asked myself what religious thinking can contribute to believers. Thus, I am going to compare the contributions of religious thinking and scientific thinking to the coping of human beings with their main existential problems.

I consider myself not only a *mathematics educator* but a *science educator* as well. The reason for that is I consider mathematics a part of science although it does not deal with the physical world. Also, all sciences (including social sciences) use mathematics in order to develop their theories. Furthermore, their structure is quite similar to the structure of mathematical theories, explicitly, the deductive structure.

As to scientific thinking, I consider it part of rational thinking. Rational thinking includes scientific thinking but is broader. Rational and scientific thinking will be discussed in Chapter 17.

Science education is an ambiguous notion. On the one hand, it denotes the act of teaching science to students, and related activities such as curriculum development, writing science textbooks, and computer software. On the other hand, it is a research discipline which investigates thought processes of students learning science and of teachers teaching science.

Science education is interdisciplinary. It borrows various tools from cognitive psychology, the psychology of problem solving, sociology, philosophy of science,

S. Vinner, *Mathematics, Education, and Other Endangered Species*,
Mathematics in Mind, https://doi.org/10.1007/978-3-319-90035-3_3

and theology. It is not supposed to make innovations in those disciplines. Thus, in this chapter I am borrowing from those disciplines; I am not trying to make any innovations.

In this chapter, I am trying to understand the thinking of religious people from the viewpoint of a secular science educator.

At this point, I would like to clarify the notion of *scientific theory* for the readers who are not familiar with it.

A scientific theory is a theory which explains the phenomena in our world and in most cases is supported by experiments or facts. A scientific theory must be refutable (Popper, 1963). On the other hand, according to Popper, there is also the notion of pseudoscience. Pseudoscience looks like science, but it is essentially irrefutable. Some examples of pseudoscience are psychoanalysis, and some parts of biological evolution.

Also, religious explanations to phenomena in our world (particularly, the creationist theory) are pseudoscientific. Religious thinking explains our world relying on religious texts. One of the most common explanations relies on the claim that things are the result of God's (the creator's) will.

I would like now to clarify some more terms. By the term *secular* I mean somebody who does not believe that God exists. For me, the term *secular* has the same meaning as *atheist*. Yet, there are some secular people who will not declare themselves atheists, but rather say they are not sure that there is no God (sometimes they will bring up one or more events that they suggest could not have happened by pure chance). In addition, the term *atheist* has, unfortunately, a negative connotation for some people. It is not politically correct. This is another reason for some to avoid declaring themselves atheists. They prefer to appear as *pantheists*. Pantheism is the belief that the universe (the totality of everything) is identical with God. This idea is due to the Jewish philosopher, Baruch Spinoza (1632–1677).

Nevertheless, since I believe that political correctness is very often the enemy of truth, I use the term *atheist* here to clarify the distinction between religious people and secular people.

The religions which I will discuss are Judaism and Christianity. From time to time, while being involved in theological discussions with my gentile friends, I remind them that Jesus was Jewish and was strongly influenced by the Jewish conception of life and the world. Another religion, Islam, about which I know very little, will be mentioned shortly and superficially.

The purpose of this chapter is to explain how scientific thinking as well as religious thinking can help us cope with our daily problems, physical and psychological. In my opinion, our main problems are suffering (physical and psychological) and mortality.

I will start with scientific thinking. Scientific thinking offers us, among other things, ways to cope with our health problems. It produced ways to deal with our physical pain and diseases. In other words, it enabled medicine to develop.

Scientific thinking also offers us explanations about our physical world from its earliest known moments. It gave us the magnificent edifice of natural sciences, which is also the basis for technology as well as medicine, just mentioned. However, science does not help us cope with our psychological suffering, nor with mortality.

My main source for religious thinking will be the Bible, which also explains how the world was created and how all that we see here came into being. So, let us start with the religious explanation to the creation, that is, with Genesis, 1 (King James translation).

At the end of 6 days of work, God was quite satisfied. "God saw everything that he had made and, behold, it was very good" (Genesis, 1, 31).

As a matter of fact, God was the only one who was satisfied. Adam and Eve were not cognitively ready to make any evaluation at that stage. Why? Because they had not eaten from the tree of knowledge of good and evil. At this point, they were not aware of their mortality. God was aware. But God did not think it was negative. However, Adam and Eve became aware of their mortality after eating from the tree of knowledge and they thought it was horrible.

This takes me directly to Ecclesiastes. It is the most pessimistic book I have ever known. It is pessimistic because of the writer's understanding that he is mortal. If I have to pick a typical quotation from Ecclesiastes, I will choose Chapter 9, 4–6: "A living dog is better off than a dead lion! The livings know that they will die, but the dead know nothing; their love, their hate and their jealousy have long since vanished."

The problem of mortality leads us to the following million-dollar question: If we eventually die, why are we born? And if we are born, what is the point of ending our lives?

Is there a way to solve this dilemma? Well, there is a simple solution. It is based on the distinction between body and soul. It is quite interesting that there is a hint to this distinction in Ecclesiastes: "And the dust returns to the ground it came from, and the spirit returns to God who gave it" (Ecclesiastes 12, 7).

However, Ecclesiastes himself, probably, was not convinced. If I am allowed to express my opinion about the man who was considered to be the wisest person in the entire world, I would claim that Ecclesiastes had a realistic way of thinking. This is, in a way, a scientific way of thinking. That is, thinking which relates to facts. The idea that there is a place to which souls go after the death of the body has no support in our world. Thus, if you don't accept this idea, what is the answer to the above million-dollar question?

The answer is quite pessimistic: Without being asked we are born and eventually we have to die. Can we carry on with our life with this message? For many of us, it is unacceptable. This leads us to the search for the meaning of life. Here is an accidental list of five references which deal with this question: Klemke (1981), Feynman (1998), Frankl (1959), Yalom (1999), and Eagleton (2007).

The following quotation was taken from a review about Eagleton's book:

> How can an English professor and literary critic [Eagleton] write a philosophical brief on the meaning of life? Well, Terry Eagleton did, and did it well. He takes us through the end of Victorian certainty and shows how Hardy and Conrad raised questions with a sense of urgency that Jane Austen never had. In the early decades of the 20th century, T.S. Eliot, Camus and Sartre brought challenges to all our values, beliefs and institutions.

> Most in the West have now accepted the view that life is an accidental evolutionary phenomenon with no intrinsic meaning. Rather than lament the loss of being part of God's design, which was often impenetrable, this clears the ground for us to give life meaning whatever we choose.
>
> A starting point is realizing that life is not a problem to be solved; if we are being practical, it really becomes more an ethical issue than metaphysical. We should be more concerned about what makes life worth living, what adds quality, depth, abundance, and intensity. Eagleton's suggestions point us towards a direction we have heard before, caring for others, compassion, becoming truly engaged. And, that is what has occupied the great novelists, poets and artists of all ages.

Unfortunately, some people do not want to follow Eagleton's altruistic recommendations. Their way to cope with the evolutionary claim that life is meaningless is by denying it. This is easily done by drinking alcohol, by smoking grass, and using other drugs.

Li-Tai-Po (701–762), a Chinese poet whose poems were used by the Czech Jewish composer Gustav Mahler, writes: "Wine is already beckoning in the golden goblet, but do not drink yet: first I'll sing you a song.... Dark is life. Dark is death."

Thus, if life is just suffering, what meaning does it have? Here is the very question about the meaning of life expressed in the movie *Hair* by Miloš Forman (1979):

"Why do I live? Why do I die?" asks the singer. And he continues:

> Where do I go? Follow the river.
> Where do I go? Follow the gulls.
> Where is the something, where is the someone, that tells me why I live and die?
> Where do I go? Follow the children.
> Where do I go? Follow their smiles.
> Is there an answer in their sweet faces that tells me why I live and die?

Hence, many people make a real effort to find a meaning to their life. Here is another examples:

This example is taken from Woody Allen's movie *Hanna and her Sisters* (1986). Mickey, Allen's protagonist, after his medical doctor tells him he does not have brain cancer, realizes all of a sudden that he is mortal. He returns to his secretary, Gail, and tells her about his discovery and about his need to find a meaning to his life. Allen wishes to share with Gail his existential problem and the following conversation takes place:

> Mickey: Do you realize on what a thread we are all hanging by?
>
> Gail: Mickey, you're off the hook. You should be celebrating.
>
> Mickey: Can you understand how meaningless everything is? ...The whole world, it's meaningless.
>
> Gail: Yeah...but you are not dying!
>
> Mickey: I'm not dying now.... I'm not gonna go today. I'm not gonna go tomorrow. But eventually, I'm going to be in that position.
>
> Gail: You're just realizing this now?
>
> Mickey: Well, I don't realize it now, I know it all the time, but, but I managed to stick it in the back of my mind...because it—it's a horrible thing to...think about! ...I gotta get some answers.

Religion can be a good answer to the lack-of-meaning problem. It is demonstrated in the above movie when Mickey asks a Catholic priest, Father Flynn, to convert him to Catholicism.

Father Flynn: Now, why do you think that you would like to convert to Catholicism?
Mickey: Well, uh, because, y—you know, I gotta have something to believe in, otherwise life is just meaningless.
Father Flynn: But why did you make the decision to choose the Catholic faith?
Mickey: First of all, because it's a very beautiful religion. It's strong and well structured.
Father Flynn: So, at the moment you don't believe in God...
Mickey: No, and I want to. I am willing to do anything. I will dye Easter eggs if it works... I need some evidence; I gotta have some proof. If I can't believe in God then I don't think life is worth living.
Father Flynn: It means making a very big leap.

A relatively easy way to deal with the mortality problem is the idea of life after death, or in Jesus' terminology, "the heavenly kingdom." However, the claim about life after death is a *pseudoscientific* claim. Namely, according to Popper's criterion, it is essentially irrefutable. This is probably the reason why there are so many people who are ready to buy it.

By inventing the Heavenly Kingdom, Jesus has killed two birds with one stone: He, supposedly, cured our fear of dying, and also promised us a huge compensation for our earthly suffering. No wonder why so many people all over the world converted to Christianity.

In order to overcome the danger that Jews too would adopt Christianity, the idea of life after death was also adopted by Judaism. However, this idea does not exist in the Jewish Bible (the Old Testament).

Now, the existential situation of human beings is as follows: We hate dying but we also hate suffering.

However, ironically enough, death can liberate us from our suffering. The question is quite simple: which emotion is unbearable? Our fear to die or our suffering?

Hence, when suffering overcomes the fear to die there is the option of suicide. Here are some examples from the Jewish history:

1. King Saul (1 Samuel, 31) committed suicide when he realized that he lost the battle with the Philistines and they were going to capture him.
2. Ahitophel (2 Samuel, 17) committed suicide after his military advice was not accepted by Absalom in his rebellion against King David. Ahitophel could not bear the humiliation and he hanged himself.
3. The defenders on Mount Masada and their families committed mass suicide when they realized the Romans were going to win the battle and as a result they would be taken to Rome as slaves.

The suicide option is also suggested by Shakespeare in Hamlet's ultimate monologue "To be or not to be":

To be, or not to be,—that is the question:—
Whether 'tis nobler in the mind to suffer
The slings and arrows of outrageous fortune,
Or to take arms against a sea of troubles,
And by opposing end them?—To die,—to sleep,—
No more; and by a sleep to say we end
The heart-ache, and the thousand natural shocks
That flesh is heir to,—'tis a consummation
Devoutly to be wish'd. To die,—to sleep;—
To sleep! perchance to dream:—ay, there's the rub;
For in that sleep of death what dreams may come,
When we have shuffled off this mortal coil,
Must give us pause. There's the respect
That makes calamity of so long life;
For who would bear the whips and scorns of time,
Th' oppressor's wrong, the proud man's contumely,
The pangs of despised love, the law's delay,
The insolence of office, and the spurns
That patient merit of th' unworthy takes,
When he himself might his quietus make
With a bare bodkin? who would fardels bear,
To grunt and sweat under a weary life,
But that the dread of something after death,—
The undiscovered country, from whose bourn
No traveler returns,—puzzles the will,
And makes us rather bear those ills we have
Than fly to others that we know not of?
Thus conscience does make cowards of us all;
And thus the native hue of resolution
Is sicklied o'er with the pale cast of thought;
And enterprise of great pitch and moment
With this regard, their currents turn awry,
And lose the name of action.

It is interesting that Shakespeare does not mention diseases as a reason for suicide. We can wonder why it is so. Is it because Shakespeare was not aware of diseases or because of literary reasons? After all, Shakespeare did not write an academic essay about reasons to commit suicide. He wrote a play.

Nowadays, many terminal patients would rather die than go on with their terrible suffering. However, being so helpless, these patients cannot commit suicide by themselves. They need help. But only in few countries all over the world it is legal to help people commit suicide. This is only an incidental remark.

Hamlet did not commit suicide because he wanted to find out what caused his father's death. Among the reasons which justify suicide in Hamlet's monologue there is despised love. Unfortunately, it is demonstrated by Ophelia in Act IV, Scene 5 of that play. A hundred and seventy years later, Werter, the hero of Goethe, also committed suicide because of despised love.

Before moving on to the next issue I would like to introduce to the readers, or to remind them, of two notions from the science of religions. The first one is *transcendental God*.

A *transcendental God* is above and beyond the world. Such a God is inaccessible, and transcends human understanding and imagination. The God of the Jewish and Christian faiths is *transcendental*. This god is also *providential*, that is, he is personal, concerned, responsive, and actively involved in the world and in human affairs.

The second notion from the science of religions that I would like to use is the notion of *immanent God*. This is a God that pervades the world. Spinoza's God is immanent; as mentioned earlier, his God is identical with the world. Spinoza rejects the conception of God as a transcendental and providential creator. According to him, God is not active in the world and in human affairs. Spinoza's immanent philosophy is considered the most radical of the early modern period.

Now, let us return to the *transcendental* and *providential* God of the Jewish and Christian faiths. This God intervenes in events in our world, and his believers expect him to reward people who follow his commands and to punish the sinners.

The situation of somebody who is convinced that he did not do anything wrong and is suffering unbearably despite that, is quite typical. There are two Biblical figures that raised this problem protesting against God: Jeremiah and Job.

In Jeremiah 12, 1 he says:

> You are always righteous, Lord, when I bring a case before you. Yet I would speak with you about your justice: Why does the way of the wicked prosper? Why do all the faithless live at ease?

Jeremiah preferred to pose his question in a general way. He did not mention that he was the one who was persecuted by the wicked. Job's complaint, on the other hand, is personal. He wanted to know why he was suffering even though he had not sinned. For him too, death was suggested as a solution to his suffering. It was suggested by his wife: "Curse God and die" (Job 2, 9), she advised. Job refused. But his general mood is the same as the mood of the prophet Jonah (Jonah 4, 30): "It is better for me to die than to live."

In Job 3, 11–13 Job says:

> Why died I not from the womb? Why did I not give up the ghost when I came out of the belly? Why did the knees prevent me? Or why the breasts that I should suck? For now should I have lain still and been quiet, I should have slept: then had I been at rest.

A similar mood is expressed in Psalm, 22: "My God, my God, why hast thou forsaken me? ...I am a worm and no man....For dogs have compassed me: The assembly of the wicked has inclosed me: they pierced my hands and my feet."

By the way, the Hebrew origin of the last sentence, "they pierced my hands and my feet," is, in fact, "like a lion, my hands and my feet" (my own literal translation). My assumption is that King James' translator, being aware that Jesus referred to Psalm 22 while suffering on the cross, and being a good Christian, decided to translate it as "they pierced my hands and my feet," because it accurately describes the crucifixion.

And indeed, for Jesus, feeling that he was abandoned by God on the cross, it was only natural to choose Psalm 22 to express his unbearable pain on the cross. Indeed, "My God, my God, why hast thou forsaken me?" were Jesus' last words on the cross before he died.

In the book of Job, God relates to Job's complaint by the following (Job, 38, 4–13): "Where were you when I laid the earth's foundation? ...Who marked off its dimensions? Who shut up the sea behind doors? ...Have you ever given orders to the morning or shown the dawn its place? ...Have you comprehended the vast expanses of the earth?" If I try to formulate God's answer to Job's complaint in our daily language it will be something like: "Who are you to question my ways of conducting the world?"

A possible answer given by the Jewish religion to Jeremiah's question as well as to Job's question is that God moves in mysterious ways. This is beautifully expressed by Job's words: "Therefore have I uttered that I understood not; things are too wonderful for me, which I knew not" (Chapter 42, 3).

In case you are still not satisfied with the difficulties that the book of Job poses to the idea of a *transcendental* and *providential* God, there is another way to cope with the book of Job, given by Jewish scholars of the second century AD. It is to claim that Job did not exist. It is only a parable.

For me, as a secular reader, the image of God that emerges from the book of Job is quite negative: God bets with Satan that Job will not lose his faith, no matter what happens to him. Thus, he lets Satan destroy Job's entire property; he lets Satan kill Job's ten children and make him sick.

Even if Job is only a parable, we, unfortunately, have met and we have heard of many people whose suffering was as bad as Job's suffering.

Moving from the biblical era to our time, the idea of a *transcendental* and *providential* God faces enormous difficulties when considering the horrible events of genocides in Auschwitz and in other places all over the world. And again, the only way for believers to justify it is by claiming that we, human beings, cannot understand the ways God operates our world.

The bottom line at this point is that the secular thinking is much simpler than the believers' thinking since it does not have to cope with all the above-mentioned problems.

Of course, this is not a call to believers to abandon their belief.

Sometimes, when I discuss these issues with my students, some of them try to convince me that God exists.

A common argument for Islamic students is that the Koran is such a beautiful book that only God could have written it. Since this is claimed in a science education course I remind my students that in science we have to examine alternative hypotheses all the time. Thus, a simple alternative hypothesis to the last one is that some human beings have such wonderful literary talent that they can write a text as beautiful as the Koran. This, of course, does not convince them. They stick to their belief and I stick to mine.

Another example of proof for the existence of God is as follows: Consider things around you, furniture, buildings, cars, etc. All of them were constructed by somebody. Is it possible that the entire world was not created by somebody? Unfortunately, this proof is not valid either. It is impossible to conclude from the fact that a table is constructed by a carpenter that the world, too, has a creator. Such an argument assumes the wanted conclusion.

In the history of the theological debates about the existence of God, there is a story about a debate between the great mathematician Euler, the religious, and Diderot, the atheist. This debate was initiated by Catherine the Great (1729–1796), the Empress of Russia, a religious lady, who was disturbed by the French atheistic movement. She invited Diderot, a philosopher and one of the main leaders of the atheistic movement to argue with Euler about the existence of God. Euler said:

$$\frac{a+b^n}{n}=x.\ \text{Therefore God exists.}$$

Diderot, whose mathematical knowledge was almost zero, realized he could not contradict Euler's argument, and therefore he returned to Paris. This anecdote tells us something about Euler. First, in order to win the debate he did not avoid unfair means.

But moreover, being a mathematician (one of the greatest in mathematical history), he knew that what he said could not be a proof that God exists. Namely, he did not need a proof in order to believe in God.

There are more proofs given for the existence of God. For instance, God appeared to me in my dream.

Again, to my science education students who claim it I say: Can you suggest an alternative interpretation to the fact that you have dreamt about God rather than the claim that God exists? Have you heard about Freud? Thus, when I try to convince my students that there is no proof for the existence of God, some of them don't believe me. Others, who developed critical thinking, understand it. Does the belief in God become weaker when somebody realizes that there is no proof for His existence? I really don't know.

However, in my opinion, the real question is not whether God exists. The real question is whether God is needed. And the answer to this question is Yes (with Capital Y). Everybody needs Him, the religious as well as the atheists. For me, as an atheist, he is a metaphor, but I am also inspired by literary texts in which God is the main figure. I am inspired by the Bible, and I am inspired by musical compositions like masses and requiems, some of which were composed by secular composers.

By saying this I have almost reached the end of this chapter. I would like to summarize it with the following comment:

As a secular science educator, I have tried to compare religious thinking to scientific thinking regarding the essential aspects of life. I have revealed my own thinking. My claims about the thinking of religious people on these issues are mere speculation.

In my talks about the above issues, I usually invite religious science educators or any religious people to express their views about the above issues. However, I have not received their reaction by now.

As a final remark to this chapter, I would like to apologize: I apologize for involving the readers with an unpleasant topic—mortality. However, we face death almost every day. We go to funerals, we write wills, and so on and so forth. I believe that most of us cope with these facts by adopting Jan Francois Vilar's famous claim: “C’est toujours les autres qui meurent.” And in English: “It is always the others who die” (Jean-François Vilar (1947–2014)). So, you see, Vilar was right. It is always the others who die.

And with this humorous statement, we usually return to our homes and families and try to enjoy our life as much as we can.

Chapter 4
Concept Formation in Mathematics: Concept Definition and Concept Image

Mathematical concepts form two categories: Primary concepts and defined concepts. For instance, in Euclidean geometry, points, straight lines, and the plane are primary concepts. An angle, a ray, and a line segment are defined concepts. That is the theoretical situation. On the other hand, there is the pedagogical question: How do we teach mathematics? And when concepts are at stake, how do we get our students to acquire the mathematical concepts that are in the curriculum? Well, all mathematics teachers, whether they are aware of learning theories or not, have their own learning theory about the way mathematical concepts are acquired by their students. Quite often, these theories are implicit. Quite often they are naive. One naive theory is as follows: I will define the concept, the students will memorize the definition and immediately after that I can assume that the concept has been acquired.

As a matter of fact, at the beginning of my mathematics education research, I decided to examine that naive learning theory. The designers of a new mathematics curriculum in Israel in the 1970s decided to establish it on a single unifying mathematical concept, the concept of function. Hence, I composed a questionnaire in which I asked students to tell me whether some correspondences which were presented are functions or not (a function is a correspondence between two non-empty (not necessarily different) sets which assigns to every element of the first set (the domain) exactly one element of the second set (the range).)

Here is the questionnaire:

1. Is there a function that assigns to each number other than 0 its square, and to 0 it assigns -1?
2. Is there a function that assigns 1 to each positive number, assigns -1 to each negative number, and assigns 0 to 0?
3. Is there a function that admits integral values for non-integral numbers and admits non-integral values for integral numbers?
4. Is there a function whose graph is as follows? (Figure 4.1)
5. In your opinion what is a function?

S. Vinner, *Mathematics, Education, and Other Endangered Species*, Mathematics in Mind, https://doi.org/10.1007/978-3-319-90035-3_4

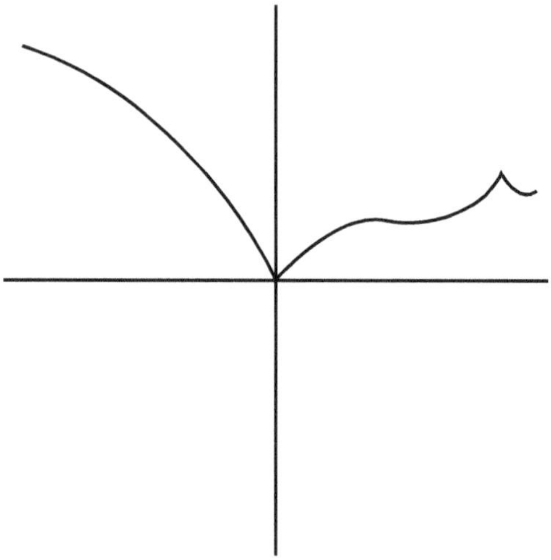

Figure 4.1 The graph that appeared in the function questionnaire

My sample included 65 students in grade 10 and 81 students in grade 11. They studied in two academically selective high schools in Jerusalem in 1978. The tenth grade students were tested several months after they had learned the chapter about functions. The 11th grade students learned this chapter when they were in the tenth grade.

If you act according to the above function definition, then the answer to all the questions is positive since to each element of the domain the correspondences assign exactly one element of the range. However, this was not what really happened. Here are some examples of ideas developed by students about the notion of function:

1. A function should be given by one rule. If two rules are given for two disjoint domains, we are concerned with two functions. This is expressed, for instance, by an answer to question 2: "No, there are three different functions. One of them gives $+1$ for all positive numbers, the second one gives -1 for all negative numbers, and the third one gives 0 to 0."
2. A function can be given by several rules relating to disjoint domains providing these domains are half lines or intervals. But a correspondence as in question 1 (a rule with one exception) is not considered a function.
3. Functions (which are not algebraic) exist only if mathematicians officially recognize them (by giving them a name or denoting them by specific symbols). This view was expressed by positive answers to question 2 like: "Yes, it is the sg function" (the "sign" function, which was taught in class). However, a student doesn't need to know about the sg function in order to answer this question positively. This is because the correspondence assigns to each element of the domain exactly one element of the range.
4. A graph of a function should be "reasonable." Many students denied the graph in question 4 to be a graph of a function because it is not regular. They claimed that a

graph of a function should be symmetrical, consistent, always increasing or always decreasing, reasonably increasing, etc.

(The reader who is interested in more detailed discussion than the above can find it in Vinner (1983).)

The results of the function questionnaire directed me to think about the cognitive mechanism that led the students to make the mistakes they made. I suggested calling it the *concept image*. At this stage of my teaching career, I already knew that definitions alone are not enough to establish concepts in the learners' minds.

The notion of *concept image* in concept formation in mathematics, as well as concept formation in everyday situations will be discussed in Chapter 9.

Chapter 5
The Misconception Fallacy, the Pseudo-conceptual and the Pseudo-analytical Behaviors in Mathematical Contexts

There was a tendency in mathematical thinking as well as in scientific thinking research to take all students' mistakes and misconceptions seriously and to suggest all kinds of thought processes that can lead to them. These mistakes were treated with such respect that some researchers even avoided calling them *mistakes* or *misconceptions*. They spoke about *alternative conceptions* (see for instance Driver (1983) and Ginsburg (1977)).

The result of this assumption was that much effort was devoted to finding cognitive interpretations for many types of behavior for which, perhaps, a different type of interpretation is more suitable. Also, much didactic effort was invested in "cognitive corrections" where perhaps a different type of correction could be more effective.

The Pseudo-conceptual Behavior

We speak a lot about *learning* situations and *problem-solving* situations, but are they really learning or problem-solving situations? They might be learning or problem-solving situations for teachers, principals, and superintendents. As a matter of fact, they should be. They are not necessarily learning situations or problem-solving situations for students. Students do not necessarily want to know something about a certain reality, whether this is a mathematical reality, a physical reality, or even a literary reality. The students are forced to learn certain topics or to solve certain problems. But nobody has control over their thoughts. Because of the *didactic contract* (Brousseau, 1988) they are not supposed to demonstrate their lack of interest in the activities imposed on them by the educational system. They will try to please the educational system with certain behaviors that are acceptable by this system. The most promising behavior that students can think of is giving the correct answer to a question that was posed to them by a teacher. Therefore, instead of *learning* or *problem-solving* processes they can try to use other processes which can

S. Vinner, *Mathematics, Education, and Other Endangered Species*,
Mathematics in Mind, https://doi.org/10.1007/978-3-319-90035-3_5

produce correct answers. I am not speaking about simple cheating. It is not cheating in the sense that you are expected to say or to do one thing and you deliberately say or do another thing. In learning, there are many cases in which the students do not know what "the real thing" is. On the other hand, they know what is accepted by most of the teachers and they know how to produce it to their teachers' satisfaction. Only a teacher who expects learning and realizes that a correct answer was obtained without any learning might be frustrated by the situation. A student, in the same situation, will probably not understand that something is wrong with his or her behavior.

As I said, a situation considered by the educational system to be a *learning* situation or a *problem-solving* situation is not necessarily such a situation for the student. Such situations I would like to call *pseudo-learning* or *pseudo-problem-solving* situations. Unfortunately, *pseudo-learning* situations (or *pseudo-problem-solving* situations) look very often like *true-learning* situations (or *true-problem-solving* situations). It is really unfortunate because as educators we would like to avoid *pseudo-learning* and *pseudo-problem-solving* situations. On the other hand, if the *pseudo* situations look very often like the true situations how would we distinguish between them? Later on I will give some examples which may help us identify the *pseudo* situations.

At this point, I would like to say a few words about the *true-learning* situations and their internal characteristics. In order to be in a *true-learning* situation, a person has to be intellectually engaged with the topic they are exposed to. It is not necessarily an emotional involvement. It is something I would call a cognitive commitment. They have to be cognitively committed to the external stimuli they are going to absorb. It does not mean that they necessarily enjoy the situation, or that they really want to be there (the contrary does not have to be true either). Being cognitively committed is a state of mind. It means that they are in a learning mode. It means that their minds are ready to assimilate some new information into their present cognitive structures or even to accommodate this structure in such a way that it will be more suitable for the assimilation process. The fact that they want it does not necessarily guarantee that learning will take place. They may want it and at the same time lack the cognitive structures which are necessary for assimilating the new topic. In such *true-learning* situations, concepts are considered, as well as the relations between the concepts, ideas in which the concepts are involved, some logical connections, and so on.

The state of mind I have described here is what I would like to call a *conceptual* mode of thinking. The result of a *conceptual* mode of thinking is a *conceptual* behavior. Using this terminology, I can claim without a serious risk that the *conceptual* mode of thinking is one of the major goals of education. Therefore, because of its major importance, we need some operational measures which will tell us whether our goal has been achieved or not.

The common operational measures are written exams, oral exams, homework assignments, class discussions, and perhaps other means as well. As a result of that, it has become possible to make a list of expected behaviors. It is not necessarily an official list but it is taken very seriously by many teachers, principals, and superintendents. It reminds us, whether we like it or not, of the behavioral goals of the *behaviorist* paradigm. It is quite ironical that even if we do not think within the *behaviorist* paradigm we cannot avoid parts of it. The point is to understand that when evaluating cognitive achievements, it is not the behavior that matters but the mental processes which produce the behavior. About these we can only speculate. The catch is as follows: one of our educational goals is to cause certain mental processes to occur in the minds of our students. In order to know whether this goal has been achieved, we have to observe the behavior of our students. But if the behavior we expect can also be produced by alternative (undesirable) mental processes then firstly, we miss our educational goal and secondly, we even might not know that we have missed it. Thus, I would like to call a behavior that might look to somebody as a *conceptual* behavior but is produced by mental processes which are not *conceptual*, namely, do not have the characteristics of *conceptual* thinking described above, *pseudo-conceptual* behavior. The mental processes that produce these *pseudo-conceptual* behaviors I would like to call *pseudo-conceptual* thought processes. At this stage, these "definitions" may look to some readers a little vague and to others a little tautological. I hope the examples I will bring later on with the discussions that accompany them, will make these notions quite clear. At this point, I would like to add only the following:

In mental processes that produce *conceptual* behaviors, words are associated with ideas, whereas in mental processes that produce *pseudo-conceptual* behaviors, words are associated with words; ideas are not involved.

There are two questions I would like to raise now:

1. Why are *pseudo-conceptual* behaviors produced, if the educational system is not officially interested in producing them?
2. How are *pseudo-conceptual* behaviors produced? In other words, what are the mental processes that lead to *pseudo-conceptual* behaviors; specifically, what are the characteristics of the *pseudo-conceptual* thought processes?

The first question is very easy to answer. It is related to the answer to the second question. Assume that there exist mental processes that are not conceptual in the above sense and can lead to a behavior considered by some of us (teachers, educators, educational administrators) as a *conceptual* behavior. Assume also that these mental processes, the *pseudo-conceptual* thought processes, are simpler, easier, and shorter than the *true-conceptual* processes. Under these circumstances, it is only reasonable to assume that many students will prefer the simpler, easier, and shorter processes to the more complicated conceptual processes. When I say *prefer* I do not want to claim that the students necessarily consider two alternatives and become convinced that one is better than the other. Some of them might not even be aware of the two alternatives. When presented with a task, they start looking for ways that will enable them to carry it out. These ways are not necessarily what the task's designers intended when they presented the task to the students. The task designers probably intended *conceptual* thought processes. The students came up with *pseudo-conceptual* thought processes. These thought processes are very often formed in a spontaneous way. They are not necessarily taught by teachers or other agents. Sometimes, they are the natural cognitive reactions to certain cognitive stimuli. Students use them without going through any reflective procedure, control procedure, or analysis.

The answer to the second question above will be given by the following examples and the discussions that accompany them.

Here are some examples of *pseudo-conceptual* behaviors:

1. In a 12th grade mathematics lesson, the teacher was giving various exercises on recursion. The following discourse took place:

   Teacher: Give a recursion rule for the sequence: $a_n = n^2$.
   Students: ...
   Teacher: What is a recursion rule?
   Student A: $a_2 - a_1$.
   Student B: $2n - 1$.
   Student C: $a_1 + a_1 = a_2$.
   Teacher: How should you express a_{n+1}?
   Student D: 1.
   Student E: n.

At this point, the teacher out of her despair gave the students the correct answer which is: $a_{n+1} = (n + 1)^2 = n^2 + 2n + 1 = a_n + 2n + 1$. Namely, $a_{n+1} = a_n + 2n + 1$.

It is quite clear that the students, demonstrating the above level of mathematical ignorance cannot gain anything from this kind of interaction. This is an example of meaningless communication in mathematics lessons. A possible way to avoid it is to eliminate such topics from the curriculum. It is related to the idea of Vygotsky (1978) about the zone of proximal development (ZPD). It is useless to teach any topic to students if it is not within their ZPD.

The answer the teacher was looking for when she asked her last question was: a_{n+1} should be expressed by means of a_n and n. The students' answers look superficially quite close to this answer. However, there was very little thinking involved in the above discussion. It was mainly blind guessing. On the other hand, if you do not know the particular topic (recursion, in this case) you might get the impression of a meaningful discussion. The characteristic of the students' reactions is as follows: The terms expressed by the teacher evoked in their minds certain associations. Since they lack understanding of the topic, they cannot examine these associations and tell whether they constitute a correct answer or not. Thus, there are two alternatives. The first one is to remain silent. The second one is to express what they have on their minds.

At least some students do not think they risk anything by telling the teacher what they have on their minds. If you do not usually practice critical thinking or you lack reflective abilities, you cannot consider the uncontrolled reaction a negative one. If you lack the mechanism to examine your associations and determine whether they make any sense in a given situation, then remaining silent is not an option. In the above behavior, the students missed only one stage: the control stage. All of us have associations when we hear or see a certain notion. We cannot control our associations, which are internal reactions to a given stimulus. However, we can control our behavior, which is the external reaction to the stimulus. A dominant feature of the *pseudo-conceptual* thought processes is the uncontrolled associations that fail to become meaningful framework for further thought processes. Hence, the recommendation is to inhibit our spontaneous reactions.

Most of us have some experience with situations in which we combine certain words or symbols and express them without knowing exactly what they mean if they mean anything. These might be test situations, professional situations, and certainly cocktail party situations. Take, for instance, the following fictitious (but typical) cocktail party conversation:

Person A: Katz is a post modernist.
Person B: I hate post modernism.
Person C: It's not my taste either but you can't ignore the impact it has on our culture.

It is quite possible that the three people involved in the above "most interesting" conversation know more or less what post modernism is. But it might well be that one of them or even all of them have no idea what it is and still talk about it in such a way that it sounds like a meaningful conversation. A very simple analysis can show how it happens. Even if the above people do not know the meaning of post modernism, they can still identify the contexts in which the term is used, and know some keywords and some stereotypical phrases associated with it. The fact that people get involved in meaningless conversations should not be criticized. It has negative, as well as positive, aspects (and let us not forget that "negative" and "positive" are relative, not absolute, notions). It can be claimed that the fact that people are involved in communication, even a meaningless one, should get some credit. This leads immediately to another example (relevant to all domains, not only mathematics) which is the "nodding behavior." You can listen to somebody, understand, and, as an indication of this understanding, nod. On the other hand, you can listen to somebody, not understand, but willing to hide this fact you can nod exactly as in the previous situation. The nodding behavior is not always a deliberately cheating behavior. Sometimes you partly understand or vaguely understand. Sometimes you believe you understand but you do not know what it means to understand in the domain you are dealing with. If a foggy landscape is the only thing you have seen in your entire life, then you will not look for clear pictures. This might be the case with many mathematics students. Sometimes they want deliberately to make a false impression on their teachers while knowing that they do not understand. Sometimes they think they understand but according to the criteria their teachers have for understanding mathematics it is not a real understanding.

An additional example: An interviewer (in an educational radio program about mathematics education): Are you claiming that Euclidean straight lines do not exist in the real world? A mathematics educator: That is correct. They do not exist in the real world. An interviewer (who has no background in mathematics or physics): These new discoveries of science and mathematics are unbelievable. Recently, physicists discovered that the world is a big chaos and now mathematicians claim that straight lines do not exist in the world.

For some listeners, this was a meaningful conversation. It looks like that. However, to the expert, who had a chance to examine the interviewer's scientific and mathematical background, it is quite clear that the interviewer had no conceptual understanding of the physical notion of chaos and of the claim that Euclidean straight lines do not exist in the real world.

Of course, at least in some situations, we will prefer a meaningful communication. I assume that educational situations are among these. But do we not make compromises when we fail to achieve our original goals?

Contrary to situations where people's intentions are to pretend they understand, there are situations in which people might believe that they think meaningfully but somebody else considers their thoughts meaningless. If meaning is a relative notion, then my claims about *pseudo-conceptual* and *true-conceptual* behaviors in such situations will face great challenges.

As a matter of fact, there are great historical controversies in philosophy, psychology, literature, and perhaps other domains where the issue of meaning is a central issue.

A given text can be considered by one person meaningful whereas the other person considers it meaningless (for instance, see Carnap (1932)). Fortunately, so I believe, the situation in mathematics learning is different. This is due to the fact that mathematicians agree on whether a certain phrase relating to a certain mathematical context is meaningful or meaningless. This situation is an immediate result of the deductive structure of mathematics. Namely, every notion is a primary notion or well defined by means of other primary or well-defined notions. The meaning of the primary notions is determined by the axioms. (In Chapter 8 I will discuss the notion of deductive system.) Thus, it is possible to claim that for mathematical contexts there exists an ultimate measure for deciding whether a given text—produced by a mathematics student in class, in a test, in a homework assignment, or in another situation—is meaningful or not. If we accept this claim, we will be able to speak about *pseudo-conceptual* and *true-conceptual* behaviors in mathematics learning contexts. This does not mean that we will be able to decide in each case whether a certain behavior is *pseudo-conceptual* or *true-conceptual.* This is because such decisions are sometimes based on speculations about thought processes that occurred in somebody's mind, and about these we may disagree.

Here is another example:

2. A class discussion after some blackboard work on an extremum problem which led the teacher to the function $y = x^2 - 16x + 60$:

   Teacher: What did we get?
   Student: A quadratic equation.

I would like to elaborate a little bit on the student's reaction. We all know that in many situations our behavior is determined by external factors, things that we see or hear. These things evoke certain associations in our minds and the associations determine, in one way or another, our reactions later on. We do not know a lot about the association mechanism. Generally speaking, we can say that an association path probably depends on the particular memory structure of the person we consider and on the particular context in which this person is. Of course, this is a very partial description but it is sufficient for the point I would like to make here. The point is that we have no control over an association path, the way it starts and the way it develops. The only thing we can control is our external reaction to the associations. It is common to speak about a control mechanism. Here, the *conceptual* thinking comes in. In some contexts, it is also called critical thinking. Identifying $y = x^2 - 16x + 60$ as a quadratic equation at first sight is a good start. If a control mechanism were activated, a suitable correction might have taken place.

The questions of what a quadratic equation is, what its general form is, how many different letters a specific quadratic equation might have, and so on, could lead our student to the conclusion that $y = x^2 - 16x + 60$ is not a quadratic equation and could have stopped his spontaneous reaction: "a quadratic equation." Additional algebraic knowledge and additional thinking of the above type could have led the student to the conclusion that $y = x^2 - 16x + 60$ is a quadratic function and not a quadratic equation.

A further analysis might lead the student to the understanding of why he made his first mistake. Unfortunately, all these never happened. The student reacted to his first association without checking his thoughts—a typical *pseudo-conceptual* behavior. It is hard to tell without further investigation into this particular student whether he has a control mechanism and whether he is capable of activating it in mathematical contexts. It is not important to the theoretical point I want to make. The point is that a *pseudo-conceptual* behavior can be caused either by a lack of a control mechanism or by a failure to activate such a mechanism. Thus, a particular instance of a *pseudo-conceptual* behavior is not an indication of somebody's intellectual ability. It is only a characterization of a particular thought process that occurred in somebody's mind. It is a mode of thinking. Of course, if the *pseudo-conceptual* behavior is quite frequent, it is an indication of the quality of thinking of the person involved. Again, it is not necessarily an indication of his or her intellectual ability. Many people have the ability of *conceptual* thinking but they rarely use it, especially in mathematical contexts.

It turns out that there is a good chance to discover *pseudo-conceptual* reactions when, without any warning, you ask your students a direct question, what is X, where X is a concept that a task they are working on deals with (as did the teacher in the first example of this chapter about recursion rules).

Here are some situations where a teacher asks a direct question:

3. An analytical geometry lesson:

   Teacher: What is the definition of a locus in analytical geometry?
   Student A: A parabola.
   Student B: The perpendicular bisector in an isosceles triangle.
   Teacher: A locus in analytical geometry is a set of points which satisfy a certain equation.

Again, as in the previous example, the lack of control is the dominant characteristic of the students' behavior. A parabola is a natural association to a locus in an analytical geometry class. However, the teacher asked for a *definition* of a locus and not for an *example* of a locus. Thus, while saying "parabola" the original question was completely ignored. In a *true-conceptual* behavior, a person is supposed to check whether the answer he or she gives is what the question required. Giving examples instead of definitions is quite typical of mathematics lessons. The reaction of the second student is even worse if we wish to rank the reactions. The student remembered something from the geometry course. The perpendicular bisector of a segment is really a locus but the student confused it with the perpendicular bisector to the basis in an isosceles triangle. This confusion can be a result of a verbal confusion or a pictorial confusion (when proving that the locus of all points which are equidistant from the ends of a given segment is the perpendicular bisector to this segment, an isosceles triangle is drawn). Again, confusions are very natural and common and should not be considered a negative phenomenon in learning. The negative phenomenon is the lack of control. Not only is the student confusing between two concepts, but his association took him away from the lesson context. (The teacher spoke about analytical geometry and the student's association took place at the context of ordinary geometry.) As I said earlier, we have no control over the way associations are evoked. However, we can control our reactions. We can examine our associations and determine whether they fit the context or not. The teacher's reaction is also quite typical for such situations in mathematics classes. The teacher just gives the correct answer. There is no attempt to relate to the wrong answers, there is no attempt to discuss them and to analyze them, there is no metacognitive discussion. Teachers should analyze students' mistakes, and explain to them why these mistakes were formed. Most students are unable to do it by themselves. Therefore, they will try to memorize what the right answer to the given question is. They will probably not understand why the other answers are wrong and why the teacher's answer is the correct one. It is just rote learning—an excellent foundation for a further *pseudo-conceptual* behavior.

Here is another situation where a teacher asks a direct question about a concept:

4. In a geometry class, the teacher is examining the knowledge of his students about certain geometrical notions.

Teacher: What is the distance between two points?
Student A: The slope.
Student B: A straight line.
Student C: A segment.
Teacher: ...The distance between two points is the length of the segment connecting the two points.
Student C: But this is exactly what I have said.

The reaction of student C is quite typical for the *pseudo-conceptual* mode of thinking. There are no clear distinctions between slightly different elements. In this case, the slightly different expressions, define essentially different mathematical entities, a geometrical entity (the segment), and a number (the length of the segment). This was a tenth grade mathematics class.

5. A direct question to a high-school graduate:

   Teacher: What is a derivative?
   Student: If $y = ax^2$ then $y' = 2x$.
   Teacher: Can you give another example?
   Student: Yes. If $y = ax^2 + bx + c$ then $y' = ax + b$.
   Teacher: But what does it mean to differentiate?
   Student: To differentiate means to divide by x.
   Teacher: And what is the derivative of $y = c$?
   Student: $y' = 0$....Here $\frac{c}{x} = 0$....The result of the division is presumably 0.
   To differentiate is presumably to divide by x.

This was an office hour event with a college student who came to my office to ask for an exemption from a calculus course. She claimed she knew calculus from previous studies. It is not clear whether she really believed it or she just wanted to skip the course. However, when asked about the central concept of the course, the student tried to construct its meaning out of her (mistaken) examples. She came up with something that was more or less coherent with her examples (dividing by x). When I "embarrassed" her with my question about $y = c$, she "invented" a system in which the result of dividing a constant by x is 0. As a matter of fact, she knew that this could not be the case. Therefore, she says with a special emphasis: The result is *presumably* 0, to differentiate is *presumably* to divide by x.

This is a typical example of a *pseudo-conceptual* thought process.

The Pseudo-analytical Behavior and Skemp's Instrumental Understanding

Richard Skemp (1976) suggested a distinction between two types of understanding in mathematics—*instrumental* understanding and *relational* understanding. In this book, I would like to avoid a discussion about understanding (such a discussion requires a special chapter relating to extremely complicated issues in philosophy and psychology). I prefer to speak about instrumental approach versus relational approach. These two approaches are also treated by other terms such as *rule oriented* versus *concept oriented*, or *procedural* versus *conceptual*. The last terms are more self-explanatory and therefore, perhaps, preferable in a community which is wider than the mathematics education community. Skemp speaks about knowing *how* and not knowing *why* versus knowing *how* and also knowing *why*. His classic example is the case of the area of a rectangle. A student can calculate the area of a rectangle just by multiplying the lengths of it sides without knowing why this product really is the area of the rectangle. (Since Skemp's classic paper, several additional papers about types of mathematical understanding have been written. Among them one can mention Backhouse (1978), Skemp (1979), and Skemp (1982). Since they are not relevant to the issues I am discussing in this book I am not elaborating on them.)

It turns out that the majority of the situations in mathematics learning are much more complex than that of the area of a rectangle problem.

Even situations that are partly similar to the area of a rectangle problem might introduce dilemmas that are not so easy to resolve. The rectangle problem is simple in a case in which the lengths of the sides are natural numbers. In this case, it is quite simple to explain why the area of the rectangle is the product of the lengths of its sides. It is less simple when one of the lengths is a fraction which is not an integer. It becomes a nuisance when the two lengths are fractions which are not integers, and it becomes a mission impossible (in school mathematics) in the case that one of the lengths or both of them are irrational.

There are many procedures in school mathematics that we teach without explaining why they obtain the results they are supposed to obtain. Take for instance the following procedures:

1. The formula for solving the general quadratic equation.
2. The formula for the area of a circle.
3. The derivative of $y = \sin x$.
4. The procedure for a graph investigation.
5. The procedure for calculating the area bounded by two graphs.

For each procedure there are, perhaps, some teachers who will try to explain to their students why it works, but many teachers will avoid it. They will avoid it not because they think it is not important, but because they believe it is too difficult for most of the students. By attempting to explain why the procedures work, these teachers might lose many students who can act well on the procedural level.

Only some of the mathematical teaching processes look like writing down the formula for solving the general quadratic equation and then solving 20 particular quadratic equations. Most situations in mathematical education are problem-solving situations, usually routine problems. In these situations, a problem is posed to the students and they are supposed to choose the solution procedure suitable for the given problem. The focus is not on why a certain procedure does what it is supposed to do. The focus is on which procedure should be chosen to solve the problem, and then on how to carry out that procedure. The question why is usually irrelevant to these situations. The intellectual challenge is the correct *selection of the solution procedure*. The student is expected to be *analytical*. It is outside the scope of this book to present a comprehensive explication of the notion *analytical*. I will use it in its common everyday meaning and I hope that the examples and the contexts in which it will appear will help the reader understand what I mean by it.

I would like to present a model for the above-mentioned mental process by means of which a procedure is selected by the student in order to solve a given mathematical problem.

A student who is asked to solve a routine problem taken from a mathematical-problem repertoire has to have the following tools:

A. A pool of solution procedures (algorithms).
B. Mental schemes by means of which the type of a given mathematical problem and its particular structure can be determined.
C. Mental schemes by means of which a solution procedure in A can be assigned to a given mathematical problem whose type and structure were previously determined.

Of course, by saying the above I am not claiming that these are all the schemes, skills, and abilities required for solving a routine mathematical problem. Everybody can point at some missing elements. It is quite reasonable, for instance, to speak especially about reading comprehension schemes and control schemes. I am not doing it because I want to simplify the model for *analytical* thought processes, as I would like to call it, in mathematical problem solving. This simplicity will help me later on when I will speak, by contrast, about *pseudo-analytical* thought processes. (As in the case of *true/pseudo-conceptual* behavior, here, too, a *true/pseudo-analytical* behavior is a result of *true/pseudo-analytical* thought processes.) As to the above pool of procedures (A), I do not want to make any claim about the way these procedures (algorithms) are represented in the mind. They can be stored in a symbolic form (like: $S = ah/2$) or in a verbal form (like: if the derivative is positive then the function is increasing). They can be fully restored or partially restored and reconstructed by various means when required.

Although the impression from the above description of the three tools is that the process is a three stage linear process, usually it is not. The type and structure of a given problem X depend on the pool of solution procedures that one has. I can decide that a given problem is an extremum problem only if I have some calculus tools to solve extremum problems. I can identify a linear programming problem only if I studied linear programming. The process of determining the type and structure of a given problem X is not a one-way process. It is a back-and-forth process. It can be a trial-and-error process. Reading comprehension plays a major role in it.

It can have parts which are guided by wild and uncontrolled association but later on control measures are taken, the outcomes are examined analytically and eventually the actions are carried out in the above-mentioned way. Therefore, the above process can be represented by the diagram in Figure 5.1.

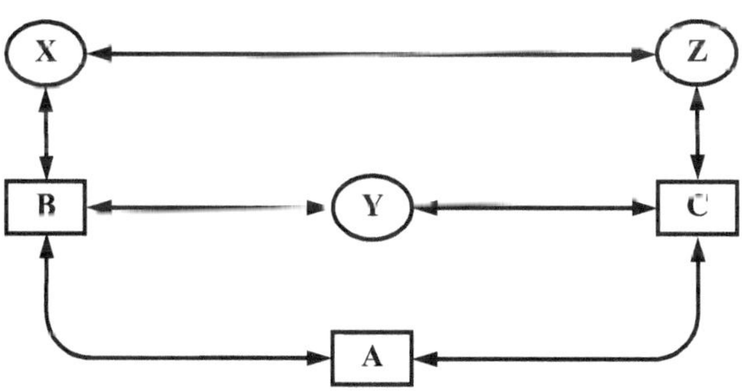

Figure 5.1 The diagram of the model for analytical selection of solution procedures. A a pool of solution procedures, B schemes that assign type and structure to a problem, C schemes that assign a solution procedure by type and structure of the problem. X a problem, Y the type and structure of the problem, Z the assigned solution procedure

The two-way arrows indicate that, as mentioned above, there are several back-and-forth sub-processes that take place before a final move is made. I would like to illustrate the above process in a schematic way through an extremely simple problem. As homage to Richard Skemp, it will be an area of a rectangle problem:

(I) Find the area of a rectangle whose sides are 7 cm and 5 cm.

A possible *analytical* line of thought for the solution of this problem is as follows:

1. This is an area problem. The area under consideration is that of a rectangle. In the given problem, the two sides of the rectangle are given. (By the end of this stage, the type of the problem has been determined: It is an area of a rectangle problem. The structure of the problem has also been determined: The two sides are given, and the task is to calculate the area.)
2. The area of a rectangle is the product of it sides (here the appropriate solution algorithm was selected).
3. The sides in this case are 7 and 5 and therefore the area is $7 \cdot 5$, namely, 35cm^2 (the algorithm which was selected has been applied to the particular problem).

The above line of thought does not necessarily describe what really happens in the mind of somebody who solves the above problem analytically. But we can consider it a schematic representation of the process.

Here is another problem which might demonstrate the complexity of the *analytical* mode of thinking even in simple cases:

(II) Find the area of a rectangle one side of which is 7 cm and the perimeter of which is 24 cm.

A possible *analytical* line of thought for the solution of this problem is as follows:

1. This is an area problem. The area under consideration is that of a rectangle. The area of a rectangle is the product of its sides. However, only one side is given in this problem. On the other hand, the perimeter is also given. The perimeter is two times the sum of the adjacent sides. By those means it is possible to find the second side of the rectangle. In this case, in order to determine the type and structure of the problem it was necessary to consider A, the pool of the solution procedures at an early stage. The type is an area of a rectangle. The structure, however, is not the common structure, since only one side is given. It is necessary to determine a strategy for finding out what the second side of the rectangle is. This requires the identification of a solution procedure for another problem: Given one side and the perimeter of a rectangle, what is the second side of this rectangle? Thus, the procedure of determining the type and structure of the given problem leads the solver to determine the type and structure of another problem, implicitly embedded in the original problem.
2. In order to calculate the second side of the rectangle, one should divide the perimeter by 2 and subtract the given side from the result. This follows from the formula: $P = 2(a + b)$, where P is the perimeter and a, b are the sides of the

rectangle. After finding the second side, the product of the two sides will be calculated in order to find the area of the rectangle.
3. $\frac{24}{2} = 12$; $12 - 7 = 5$; $7 \cdot 5 = 35$. The area is 35cm^2.

The above line of thought indicates the complexity of the thought processes involved even in simple, but less typical word problems. It also emphasizes what I claimed earlier that the above model is very schematic. What really happens is that during the analysis of the problem some *sub-problems* are identified, analyzed, and solved. Thus, the process described in the model is also applied to the *sub-problems*. I did not bother to construct a more detailed model for the solution procedure because I wanted to avoid extremely complicated diagrams.

The examples above illustrate the *analytical* mode of thinking and behavior in problem solving. I would like to describe now what I would like to call the *pseudo-analytical* mode of thinking and behavior. This mode is a result of the common teaching-learning style, and, needless to say, of human nature.

Usually, when a mathematical topic is presented to students (like the area of a rectangle) a set of problems follow. The goal is probably to illustrate the use of the information just presented to the students (like $S = ab$), and also to help the students memorize and internalize that information. In the common style of teaching and learning, the students notice immediately that this activity (solving typical problems of the topic under consideration) is the heart of school activities. It is done during class hours, it is given as homework assignment and it is a part of the common exams that determine the future of all students in school and outside school. Of course, the students do not care much about the nature of their thought processes. They care about the answers they give, whether they are accepted or not, whether they are right or not. The most important thing for the students is to give the right answer. My assumption is that, usually, the preferable procedure to achieve a goal is the minimal effort procedure. This assumption can be called the minimal effort principle. In our case (the area of a rectangle), the teachers can elaborate a lot about the formula $S = ab$. They can demonstrate again and again an analysis of a given area problem. However, for the students the bottom line is that in order to obtain the right answer in the typical examples, one should multiply the two numbers mentioned in the problem. Thus, the method of solving an area problem given after the rule for computing an area of a rectangle was given to the students is as follows: identify the two numbers in the problem and multiply them. Since in the future, area problems might be given in broader contexts, the students will try to recall the typical area problem to which a given problem is similar. Then, they will apply the solution procedure for the typical problem. In our case, they will multiply the two numbers given in the problem.

Thus, a possible line of thought for solving problem (I) above is as follows:

1. Problem (I) is similar to problems we solved when we studied about the area of a rectangle.
2. These problems were solved by multiplying the two numbers given in the problems.

3. The two numbers in this case are 7 and 5 and therefore the result is $7 \cdot 5 = 35$. A more satisfactory answer is 35cm^2.

This line of thought I would like to call *pseudo-analytical*. In the case of problem (I), it even leads to a correct answer.

Now I would like to present a model for the *pseudo-analytical* thought process by means of which a procedure is selected in order to solve a given mathematical problem.

A student who solves a problem using the *pseudo-analytical* process has to have the following tools:

A′. A pool of typical problems and their solution procedures.
B′. Mental schemes by means of which a similarity of a given problem to one of the problems in A′ can be determined.

The similarity found in B′ is very often a *superficial similarity*, based on a *fuzzy memory* of that problem in A′.

When a problem X is posed to a student who is in the *pseudo-analytical* mode the following process will take place:

1. B′ will be activated by X. It will act on A′ in order to identify a problem Y which is similar to X.
2. The solution procedure for Y will be applied to X.

Note that the *pseudo-analytical* process is much simpler and much shorter than the *analytical* process. I am not claiming that somebody consciously decides to choose the *pseudo-analytical* process and to avoid the analytical process. In order to do that, one must have reflective abilities and be in a reflective mode. Usually, young children do not have reflective abilities.

On the other hand, people who have reflective abilities do not necessarily use them in every situation. Thus, it is quite possible that even professional mathematicians will act in a *pseudo-conceptual/analytical* mode in certain situations. The most characteristic feature of the *pseudo-conceptual/analytical* behavior is the lack of control procedures. The person is responding to his or her spontaneous associations without a conscious attempt to examine them. The option of inhibition was forgotten. The moment a result is obtained there are no additional procedures to check the correctness of the answer. The model for the *pseudo-analytical* process for solving a problem can be illustrated by the diagram in Figure 5.2.

Assume now that problem (II) above is posed to somebody who happens to solve it using a *pseudo-analytical* process. A possible line of thought for this question can be:

1. The problem looks like an area of a rectangle problem.
2. In such problems, the answer was obtained by multiplying the two numbers in the problem.
3. The two numbers in this problem are 24 and 7. Therefore, the answer is $24 \cdot 7$, namely, 168.

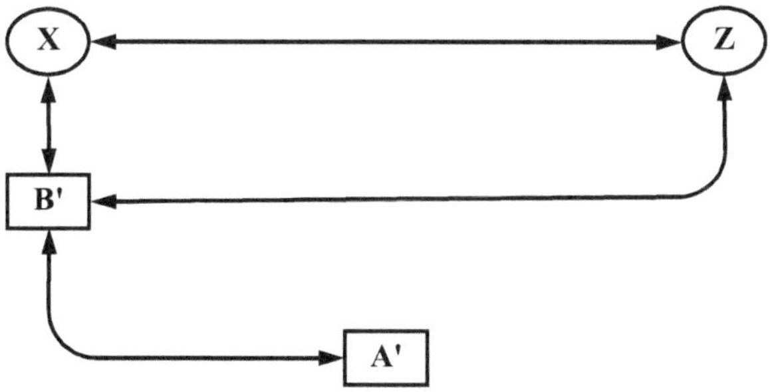

Figure 5.2 The diagram of the model for *pseudo-analytical* selection of solution procedures. A′ a pool of problems and their solution procedures, B′ schemes that determine similarity of the problem to a problem in A′. X a problem, Z the similar problem in A′ and its solution procedure

In this case, contrary to the *pseudo-analytical* solution for problem (I), the fact that the *pseudo-analytical* process was used becomes immediately obvious.

It is also obvious that our educational goal is not to encourage the *pseudo-analytical* behavior. It is impossible and perhaps undesirable to eliminate this behavior. But, as a rule, our goal is to encourage *true-analytical* behavior. One way to encourage the students not to choose *pseudo-analytical* behavior is the sanction method. Specifically, we should pose problems to them such that trying to solve them in the *pseudo-analytical* mode will result in an erroneous answer. Problem (II) is such a problem. However, if (II) were given in a common exam, some students (or even parents) might raise the fairness issue. Since this issue is, perhaps, the most sensitive issue in the educational system, teachers try to avoid it as much as possible. Thus, the chance to minimize the number of situations in which a student will use the *pseudo-analytical* mode becomes quite small.

The Pseudo-conceptual, the Pseudo-analytical, and How They Are Related

I am using two different notions in order to distinguish between two different contexts. In the first context, where I use the notion *pseudo-conceptual*, the focus is on concepts. In order to act correctly, one should relate to the concepts involved in the context. The emphasis is on the meaning of the words, the symbols, and the mathematical notions. The second context, where I use the notion *pseudo-analytical*, is a problem-solving context. *Analytical* thought processes in this context were described in the previous section.

Nevertheless, there are some contexts in which both the analytical and the conceptual (or the *pseudo-analytical* and the *pseudo-conceptual*) modes are involved. Therefore, it might seem that the distinction between the two notions is not clear. In spite of that I think that it is preferable to keep the two different notions. There are contexts which are essentially problem-solving contexts. In these contexts, the main activity is to analyze the situation and to find the suitable solution procedure.

It is true that part of the analysis may be conceptual; the control procedures in particular may involve conceptual elements. However, if the main process is a problem-solving process, I will speak about analytical or *pseudo-analytical* thought processes and behaviors. On the other hand, if the main process focuses on concepts, their interrelations, and their application, I will speak about conceptual or *pseudo-conceptual* thought processes and behaviors. The differences will be illustrated in the next sections, where I will describe some additional examples of *pseudo-conceptual* and *pseudo-analytical* processes and try to explain how they are formed.

An Example of Pseudo-analytical Process: A Superficial Similarity and a Fuzzy Memory

This example is taken from the quadratic equation chapter. A similar problem appears in Matz (1982), but since the analysis there is somehow different from the one based upon the notion of *pseudo-analytical* I would like to discuss it here at length. Sometimes, a quadratic equation is given as a product of two linear polynomials on one side and zero on the other side of the equality sign. Later on, the students are taught to factorize the left-hand side of a quadratic equation given in its *canonical form*, $ax^2 + bx + c = 0$. The students can figure out the solutions of the quadratic equation from the factorization. Usually, on the class blackboard it looks like that:

$$x^2 - 5x + 6 = 0$$

$$(x - 2)(x - 3) = 0$$

$$x - 2 = 0 \quad \text{or} \quad x - 3 = 0$$

$$x = 2 \quad \text{or} \quad x = 3$$

The mathematical justification for each step is usually not written and very often not even mentioned orally.

The students carry out an algorithm in an instrumental approach. They focus on the *how* and not on the *why*. Consider now the following quadratic equation: $x^2 - 5x + 6 = 2$.

When posed to students in an average class, some of them solve it as follows:

$$(x-2)(x-3)=2$$
$$x-2=2 \quad \text{or} \quad x-3=2$$
$$x=4 \quad \text{or} \quad x=5$$

Surprisingly enough, $x = 4$ is a solution of the equation, so if this were an exam, it even might have received partial credit. The explanation of the students' procedure is quite simple in this case. There is an obvious *superficial similarity* between $x^2 - 5x + 6 = 0$ and $x^2 - 5x + 6 = 2$. The students did not remember that the factorization procedure can be applied only in the cases where there is a zero on the right-hand side of the equation. It is clear that they did not have a relational approach to the factorization solution procedure. Had they remembered it correctly, even without understanding its justification, they would not have used it because the conditions for using it are not fulfilled. Thus the identification of a *superficial similarity*, without the ability to realize that the equations which were solved in their previous experience are structurally different than the newly given equation, led the students to use an unsuitable solution procedure. The procedure the students used can be regarded as an imitation of the original procedure for solving the equation: $x^2 - 5x + 6 = 0$.

But it is quite possible that the type of the original equation was stored in the students' minds as $x^2 - 5x + 6 = k$ (where k is any number). This can be considered as a *fuzzy memory* event. The rest of the procedure in this case will be:

$$(x-2)(x-3)=k$$
$$x-2=k \quad \text{or} \quad x-3=k$$
$$x=k+2 \quad \text{or} \quad x=k+3$$

If no conceptual analysis of the procedure is done, the students cannot realize that the procedure they are using is senseless. They also do not check their answers. Had they understood the notion of a solution for an equation, they could substitute their numerical results for x and find out that one "solution" does not fulfill the condition it is supposed to fulfill. Probably, none of these were carried out during the solution process or after it.

I would like to emphasize two important cognitive faculties. The first one is the ability to identify similarities. The second one is the imitation ability. Without these abilities, some important intellectual activities would not even occur. The problem is not that these abilities were not involved in the above solution procedure. The problem is that some additional thought processes, the analytical and the conceptual, were missing.

An Additional Example of Pseudo-analytical Process: A Superficial Similarity and a Fuzzy Memory

The following problem was given to 50,000 high-school students in Illinois in a "simulation" of a university mathematics placement exam: $-x^{-1} + x = ?$

It was a multiple choice exam. Out of the 50,000 students only 21% got the correct answer $\frac{x^2-1}{x}$. The distracter $2x$ was chosen by 27%. It was chosen three times more often than the correct response in the four lower quintiles (Peressini & Morris, 1983).

It is clear that this is not a random result. It is hard to tell what led so many students to this distracter because the students were not asked to explain their choice. However, the explanation that I would like to suggest is as follows: The algebraic expression $-x^{-1}$ reminds some students of the algebraic expression $--x$. Therefore, for them $-x^{-1} = --x$. In both expressions, the minus sign appears twice. Since $--x = x$ then the conclusion is that $-x^{-1} = x$. Therefore $-x^{-1} + x = x + x = 2x$. If this thought process really took place, then the failure is a *fuzzy memory* failure. Only if your memory is fuzzy or vague you can identify $-x^{-1}$ as $--x$. It indicates that there does not exist somewhere in your mind a clear picture of $--x$, or $-x^{-1}$, or both. There is, perhaps, a vague picture of "an expression with two minus signs." If this is what you really remember, then $-x^{-1}$ can be easily identified as $--x$. It is like remembering that there is a tree at your friend's backyard. However, you do not remember exactly what kind of tree it is. In a certain context, if you are told that one of the trees in a given list (or in a given picture) is your friend's tree you can easily make serious identification mistakes (similar identification mistakes occur quite frequently in court). The moment the wrong identification is made there is nothing which can prevent the final conclusion: $-x^{-1} + x = 2x$. The above wrong identification is probably a result of the confusion between x^{-1} and $-x$. This confusion is, perhaps, not only a visual confusion. It may also be a conceptual confusion. Conceptually, x^{-1} and $-x$ are not similar. The first one is the multiplicative inverse and the second one is the additive inverse of x. Hence, this is a *superficial similarity*.

An Additional Pseudo-analytical Process: A Superficial Similarity and a Fuzzy Memory

Consider the following example which was taken from a computer-based learning session. A 9-year-old boy was solving some exercises posed to him on the computer screen. I was sitting beside the boy and interfered whenever the pupil confronted serious difficulties.

Computer: Which number is greater than 4 by 7?
Pupil: 3.
Computer: You are wrong.

I: What did you do? Explain it to me!
Pupil: I calculated the difference.
I: The difference between what?
Pupil: The difference between 7 and 4.
I: What was the question?
Pupil: ...Which number is greater than 4...
I: By 7.
Pupil: I do not understand it.
I: ...Do you know, perhaps, which number is greater than 1 by 2 ?
Pupil: Of course!
I: Which?
Pupil: 1.
I: Again you are calculating the difference....What does it mean that 5 is greater than 3 by 2?
Pupil: ...Plus 4!
I: ...Alright, try it!
Pupil: I'll make it 11.
Computer: This time you are right.

When analyzing the above dialogue, it is almost impossible not to conclude that the pupil does not understand the meaning of the expression "a number which is greater than y by x." (This was expressed explicitly by the statement: "I do not understand it.") On the other hand, he is probably familiar with an expression similar to the above expression which is: "By what number is x greater than y?" He also knows that in order to answer the last question he has to subtract y from x. This is probably the reason for the calculations $7 - 4$ and $2 - 1$. Thus, the pupil relies on some *fuzzy memories* and on *superficial similarities*, believing that the given question is similar to questions he solved in the past by subtraction. These questions were of the type: "By what number is x greater than y?" The strategy for solving these questions was the subtraction strategy, that is, $x - y$. The moment the pupil realizes that this strategy does not work, he looks for another strategy. There is no attempt to understand *why*. There is only an attempt to find out *how*. This attitude will probably not lead the pupil to the expected *analytical* behavior. The analysis required by the *analytical* behavior is very often based on meaning inquiry (what does it mean that a given number is greater than x by y?). In the above session, however, the moment the right answer is obtained there is no motivation to learn anything that seems irrelevant to the situation. As a matter of fact, nothing is relevant besides the right answer.

The question "Which number is greater than 4 by 7?" stimulated in the child's mind the subtraction strategy. He was stuck with it, lacking the *conceptual* tools to get out of it. Using the addition strategy at the end does not indicate that he really understood why it was the right one for solving the question which was posed to him by the computer. It might have been pure guessing.

The above pupil's behavior can also be related to Nesher and Teubal (1975). The issue there was that pupils look for verbal cues in order to decide about the operation needed to solve the exercise.

More Examples of Pseudo-analytical Processes: Superficial Similarities and Fuzzy Memories

The following question was given to a college calculus class after they were reminded the concept of function:

"Are there two different numbers such that the function $y = x^2 + x + 1$ assigns to both the same value?"

Forty percent of the students ($N = 30$) wrote the equation $x^2 + x + 1 = 0$, tried to solve it, and claimed that the answer to the question is negative because the equation has no solution. It is quite clear that these students did not understand the question. It is not a common question after all. Its syntactical structure is, perhaps, not so simple. It is expected, however, that a college student will make an effort to comprehend the text before applying any solution procedure. Such an effort was not made by these students. They were led by the "quadratic-equation reflex" which is: Whenever you identify a quadratic form solve it by means of the solution formula. Ironically, the stimulus to the "quadratic-equation reflex" was not a quadratic equation but something which looks like a quadratic equation ($y = x^2 + x + 1$). This does not disturb the students to construct from it a "real" quadratic equation ($x^2 + x + 1 = 0$) and to give an answer which seems to be implied by the fact that the equation has no solution (no solution means a negative answer). On the surface, things look quite alright.

Another example of *pseudo-analytical* behavior has been exposed by the following humorous word problem (Verschaffel, 1999): "There are 26 sheep and 10 goats on the ship. How old is the ship's captain?"

More than 60% of students attempted to solve the problem by adding the two given numbers. Hence, their answer was 36. An expected *true-analytical* behavior should imply the following reaction: There cannot be any connection what so ever between the numbers of sheep and goats on the ship and the age of the captain. However, the culture of questions and answers in mathematics class rooms encourage the students to do some manipulations on the numbers in the word problem in order to give an answer. The students are not encouraged to criticize the questions and to claim that with information given in the problem it is impossible to answer the question.

The Pseudo-conceptual and Pseudo-analytical Modes of Thinking and Mathematical Misconceptions

My claim is that the *pseudo-conceptual* and the *pseudo-analytical* behaviors are very common in mathematics learning. They are much more common than most mathematics teachers believe. I also think that they deserve the same research attention as misconceptions.

The *pseudo-conceptual* and *pseudo-analytical* modes of thinking and misconceptions are closely related. Very often it is hard to draw a line between them. This fact, however, is not a reason to consider the *pseudo-conceptual/analytical* modes a part of misconception theory. The difference between the *pseudo-conceptual/analytical* modes and misconceptions is in the cognitive dimension. The way I see it, misconceptions occur in conceptual frameworks. The *pseudo-conceptual/analytical* modes are outside the conceptual frameworks.

A misconception is based on a belief about a certain (mathematical) situation. *Pseudo-conceptual* or *pseudo-analytical* thought processes are based on the belief that a certain act will lead to an answer which will be accepted by society or which will impress society (the mathematics teachers in our case). In other words, I am dealing with the difference between the belief that a statement X is true and the belief that a statement X will be credited by a person Y who is supposed to evaluate it.

A misconception is a result of cognitive involvement. It is a result of cognitive efforts. These efforts led to a wrong idea. Nevertheless, it is a cognitive belief about "the world." On the other hand, when *pseudo-conceptual/analytical* behaviors occur there is no cognitive involvement. The person is looking for a satisfactory reaction to a certain stimulus while cognitive issues do not play a role. The thought processes are guided, in addition to other things, by uncontrolled associations and *superficial similarities*. These can lead to superficial analogies and superficial generalizations. Thus, the distinction between the *pseudo-conceptual/analytical* modes and misconceptions is a theoretical one. Sometimes it will be impossible to determine whether a particular event is a result of a misconception or a result of a *pseudo-conceptual/analytical* thought process. It depends on what really happened in somebody's mind, something that we can never know for sure.

If you take some of the examples in Matz's classic paper (1982), mentioned already in a previous section of this chapter, you can consider them either misconceptions or *pseudo-conceptual/analytical*, depending on what happened in the student's mind. For instance, the case of: $(A + B)^2 = A^2 + B^2$ (p. 29 there).

Matz suggests that students, after being exposed to $(A + B)C = AC + BC$ and $(A - B)C = AC - BC$, are led to believe that it is true for any operation, and thus, they form in their minds an abstract schema $(X \square Y) \triangle Z = (X \triangle Z) \square (Y \triangle Z)$ (the symbols $\square$ and $\triangle$ denote operations). This schema is triggered when they solve "simplify: $(A + B)^2$", and they write $(A + B)^2 = A^2 + B^2$.

If this were the line of thought it is possible to consider it a misconception. It contains a cognitive step (Though, one can argue that this step was not followed by a control procedure which the student, undoubtedly, saw in class many times, namely, substituting numbers for the letters on both sides, calculating, and comparing the results. When focusing on this missing step by itself one can argue that it was a *pseudo-analytical* thought process.)

On the other hand, the following line of thought could also take place:

There is a rule $(A + B)C = AC + BC$. Our present exercise is quite similar. Therefore I'll use: $(A + B)^2 = A^2 + B^2$. If this were the line of thought I would, without hesitation, consider this a *pseudo-analytical* thought process. Matz claims (p. 25) that "errors are the result of reasonable, although unsuccessful, attempts to adapt previously acquired knowledge to a new situation." My focus is on "reasonable attempts." What is a reasonable attempt? What is reasonable? Somebody is supposed to decide what reasonable is. Are there objective criteria for reasonable thought processes? It is hard for me to accept that every error is "a result of reasonable, although unsuccessful, attempts to adapt previously acquired knowledge to a new situation." Those errors which are not, I would like to consider *pseudo-conceptual* or *pseudo-analytical*. There is no point in focusing on borderline cases. The focus should be on the typical cases. The path from the *pseudo-conceptual* to the conceptual is in many cases "continuous." Borderline cases will, therefore, remain undecidable.

In any case, as with misconceptions where researchers try to explain their origins, researchers should also try to explain the origins of the *pseudo-conceptual* and the *pseudo-analytical* behaviors, how they are formed and what the principles which guide them are. The research in this case, as educational research, has two goals. The first one is to understand the student's behavior. This is the basic research approach. The second goal is a result of the first one. Only after understanding our students' behavior it is possible to suggest ways to replace them through the course of learning by *true-conceptual* and *true-analytical* thought processes. In this chapter, however, I will not make any didactic suggestions. I would only like to emphasize that the treatment given to *pseudo-conceptual/analytical* thought processes should be different than the treatment given to misconceptions.

Verbal Cues

As I already said several times, *pseudo-conceptual* and *pseudo-analytical* thought processes start very often with spontaneous, natural but uncontrolled associations. In many cases, these associations are a response to a verbal stimulus. This implies that verbal stimuli are used as *cues* to solve problems. This is, in fact, the idea in Nesher and Teubal (1975), a study of word-problem solving in third grade, which I have already mentioned earlier.

Using my terminology, the research methodology there was to design questions that would reveal *pseudo-analytical* strategies in solving word problems. Take, for instance, the following question (p. 51): "On Monday the milkman brought 7 bottles of milk. That was 4 bottles less than he brought on Sunday. How many bottles did he bring on Sunday?" A child in a *true-analytical* mode of thinking will carefully read the question, understand the situation and will realize that what is needed in order to answer the question is to add 7 and 4. On the other hand, a child who is in a *pseudo-analytical* mode will identify the two numbers, 4 and 7, will note the word *less* which is usually associated with subtraction and will subtract 4 (the smaller number) from 7. In fact, this is what many children do in grades 3 and 4. Now, a different description of the above situation is presented to these children: "The milkman brought on Sunday 4 bottles of milk more than on Monday. On Monday he brought 7 bottles. How many bottles did he bring on Sunday?" Being still in the *pseudo-*

analytical mode, these children will identify the numbers 4 and 7, will note the word *more* which is usually associated with addition and will add 4 and 7. This time they will get the correct answer and nobody will know that they acted in a *pseudo-analytical* mode. Hence, a systematic research of *verbal cues* will help identify and also overcome *pseudo-analytical* behaviors in situations we do not want them. The research is needed, of course, not only in the domain of third grade word problems but in all grade levels, including college level, and in all central topics: functions, derivatives, continuity, limits, graphs and many more.

As a matter of fact, many research findings about misconceptions, implicit models and problem-solving strategies in all levels of mathematics learning can be reinterpreted in the *pseudo-conceptual*, and the *pseudo-analytical* theoretical frameworks suggested here.

Dealing with the Pseudo-conceptual and Pseudo-analytical Behaviors

It is impossible and also undesirable to eliminate the *pseudo-conceptual* and the *pseudo-analytical* behavior. We should only control them.

Pseudo-conceptual and *pseudo-analytical* thought processes may be classified according to the state of mind (of the person under consideration) during the processes:

1. They are not aware of their *pseudo-conceptual* or *pseudo-analytical* thought processes. They are reacting to a stimulus in a spontaneous manner and following the line of thought which has been formed in their minds. However, they have a control mechanism. The moment they activate the control mechanism they can examine analytically the results of their *pseudo-conceptual* or *pseudo-analytical* thought processes and determine whether they make sense or not.
2. As in the previous case, they are not aware of their *pseudo-conceptual* or *pseudo-analytical* thought processes but they do not have a control mechanism which can tell them whether they acted in a *pseudo-conceptual* mode or not. Namely, they do not have analytical abilities for the particular situation.
3. They are in the poseur situation. They know that they do not know but they want to make the impression that they do know, just in order to get some credit (sometimes an academic credit, sometimes a social credit).

As educators we would like to help our students construct the *true-conceptual* and the *true-analytical* modes of thinking. Nevertheless, it is impossible to ignore the fact that the *pseudo-conceptual* and the *pseudo-analytical* modes of thinking are spontaneous, natural and, in many cases, also effective. Therefore, the appropriate way to encourage students to use the *conceptual* and the *analytical* modes is by helping them construct awareness. We should try our best to get our students cognitively involved. The lack of cognitive involvement is one of the reasons for the common use of the *pseudo-conceptual* mode. Unlike immoral behaviors, we do not want to eliminate *pseudo-conceptual* and *pseudo-analytical* behaviors. Very often we need *pseudo-conceptual* behavior as a starter to a thought process in which some conceptual elements will appear later. It is even possible that the *pseudo-conceptual* mode precedes the conceptual mode developmentally. If we look at the way babies learn their mother tongue we see that first they learn to identify the appropriate context of certain words and then they use these words in these contexts. Only later on we can assume that certain meanings are associated with certain words. What is true about learning the mother tongue might be partially true about learning mathematics at later stages.

The *pseudo-conceptual* and the *pseudo-analytical* modes, sometimes, are survival behaviors. It does not matter whether they occur in an exam or in a cocktail party. It is pointless to convince somebody to abandon a survival behavior.

Also, particular events of *pseudo-conceptual* behavior are not an indication of lack of mathematical talent. They might be a result of concentration problems, fatigue or temporary lack of motivation. They can occur to professional mathematicians as well as to all kinds of mathematics students. In an interview with an outstanding mathematician the interviewer posed to the mathematician the following question: There are 200 trees in an avenue. The distance between any two trees is 2 meters. What is the length of the avenue? The outstanding mathematician replied immediately 400. He was tempted to multiply 200 by 2. However, this answer is wrong. If you are not convinced take the simplest case where the avenue has only two trees and the distance between them is 2 meters.

Thus, we are in an extremely complicated situation. Not only is it quite hard, sometimes, to identify *pseudo-conceptual* and *pseudo-analytical* behaviors, but we should also remain ambivalent about them. It is pointless to declare a total war against them. On the other hand, in many cases, we do not want to encourage them. It seems that the formation of control mechanisms in somebody's mind might help a lot in this direction. This can be done by encouraging people toward reflective thinking. I have already spoken about it in the section about reflection in the preface. Unfortunately, not everybody is capable of reflective thinking, and those who are, don't necessarily like it, certainly not all the time. Thus, the best we can expect in this context is to teach ourselves and our students how to identify our *pseudo-conceptual* and our *pseudo-analytical* thought processes and how to control them when control is needed.

The bottom line is that we should not declare war against the *pseudo-conceptual/analytical* behavior. Everybody can find himself or herself in *pseudo-conceptual* behavior situations. For instance, a situation of my own experience. Being a foreigner in the USA, and not a native English speaker—quite often I do not understand what people are saying to me. There are situations in which I ask them to explain to me what I have failed to understand. On the other hand there are situations in which it is quite embarrassing to ask for an explanation. Such a situation occurred to me at a Subway restaurant.

The person who prepared the sub asked me several questions about the kind of bread that I would like to have, what sauce, and what flavor. He named the kinds of bread, the sauces and the flavors. However, the names did not mean anything to me. It seemed to me inappropriate to ask for explanations because there were many people on line after me. So, I repeated some of the words of the Subway person. Unfortunately, I did not get the sub I hoped to have, but I did get satisfaction from the fact that I did not cause a delay on the line behind me.

A final humoristic example: Sometimes adults ask kids or even other adults: "What is heavier, 1 kilo of iron or 1 kilo of feathers?" There are some people who fall in that trap and reply: "1 kilo of iron." It turns out that the answer is typical of a *pseudo-conceptual* line of thought. The person reacts to the stimulus of feather and iron. Of course, iron is heavier than feathers if you are talking about the density of the two materials. However, the question asked to compare between 1 kilo of iron and 1 kilo of feathers. The person did not pay attention to that. He or she was not in an *analytical* mode of thinking and thus gave the wrong answer.

Two Literary Examples

Some classic writers who wrote about aspects of education and culture noticed the *pseudo-conceptual* behavior long time ago. Thus, it is a well-known phenomenon. In this book I only suggested a terminology and a cognitive analysis.

I would like to point very briefly at two great examples. The first one is Molière, *Le Bourgeois Gentilhomme* (1670). The English translator called it *The would-be Gentleman*, which is already an interpretation. The way I understand it, a "would-be gentleman" is somebody who strongly desires to become a gentleman but who is not very successful. This leads us to the category of the pseudo-phenomena, the pseudo-gentleman, in this case. In the following two excerpts pseudo-education is wonderfully illustrated. In the first one, the student (Mr. Jourdain) is talking to his teacher (the philosopher).

Mr. Jourdain: . . .I am in love with a lady of quality and I want you to help me write her a little note I can let fall at her feet.
Philosopher: Very well.
Mr. Jourdain: That's the correct thing to do, isn't it?
Philosopher: Certainly. You want it in verse, no doubt?
Mr. Jourdain: No. No. None of your verse for me.
Philosopher: You want it in prose then?
Mr. Jourdain: No. I don't want it in either.
Philosopher: But it must be one or the other.
Mr. Jourdain: Why?
Philosopher: Because, my dear sir, if you want to express yourself at all there's only verse or prose for it.
Mr. Jourdain: Only prose or verse for it?
Philosopher: That's all, sir. Whatever isn't prose is verse and anything that isn't verse is prose.
Mr. Jourdain: And talking, as I am now, which is that?
Philosopher: That is prose.

Mr. Jourdain: You mean to say that when I say "Nicole, fetch me my slippers" or "give me my night-cap" that's prose?

Philosopher: Certainly, sir.

Mr. Jourdain: Well, my goodness! Here I have been talking prose for forty years and never known it.

Till a certain point, the above conversation sounds like meaningful communication ("You want it in verse no doubt?" "No. No. None of your verse for me."). Only at a crucial point it turns out that the student has no idea of the topic discussed ("No. I don't want it in either."). This is quite similar to the excerpts in previous sections where mathematical subjects were discussed by real teachers and students. There is also an implicit criticism about teaching in the above dialogue, and this is quite relevant to our concern about the *pseudo-conceptual* mode. Because of the impossible conditions for learning (in this case, the impossible student), the teacher is forced to teach in an inadequate way. I am quite sure that Molière was also critical about "the significance of that particular lesson" to human culture. But even if we ignore this, it is crucial to remember that the distinction between verse and prose is a distinction between literary forms. It is taken from the theory of literature. It is quite ridiculous to apply it to everyday language. The philosopher did it as a kind of didactic simplification. The student, lacking the required background to assimilate this (here I am using cognitive terminology with which Molière was probably unfamiliar) handled it in the *pseudo-conceptual* mode. This is illustrated in an even more ridiculous way when the student becomes a teacher. This is also relevant to my claim that certain modes of teaching encourage the *pseudo-conceptual* behavior. In the next excerpt, the student is talking to his wife.

Mr. Jourdain: …I want to improve my mind and learn to hold my own among civilized people.…For example, do you know what you are doing—what you are talking at this very moment?

Mrs. Jourdain: I'm talking plain common sense—you ought to be mending your ways.

Mr. Jourdain: That's not what I mean. What I am asking is what sort of speech are you using?

Mrs. Jourdain: Speech. I'm not making a speech. But what I'm saying makes sense and that's more than can be said for your goings on.

Mr. Jourdain: I'm not talking about that. I'm asking what I am talking now. The words I am using—what are they?

Mrs. Jourdain: Stuff and nonsense!

Mr. Jourdain: Not at all. The words we are both using. What are they?

Mrs. Jourdain: What on earth are they?

Mr. Jourdain: What are they called?

Mrs. Jourdain: Call them what you like.

Mr. Jourdain: They are prose, you ignorant creature!

The declaration about improving the mind, when taken away from the above context, may sound like an excerpt from National Council of Teachers of Mathematics, *Curriculum and Evaluation Standards for School Mathematics* (NCTM, 1989). The rest of the dialogue may be considered an illustration for the constructivist approach to teaching or learning by discovery. Whatever our associations are, the combination of a teacher who does not understand the purpose of teaching a certain topic and a student who does not believe the topic has any relevance to her is, unfortunately, quite familiar to us, as mathematics teachers.

The second literary illustration of *pseudo-conceptual* behavior is taken from Ionesco (1951), *The Lesson*. As I understand it, the play demonstrates how meaningless human communication may be and how art and science are used as pointless means of intellectual activity. The theatrical framework of the play is a private lesson given by a Professor (a man in his fifties) and a young girl, 18 years old. Almost all domains of human culture are represented in this lesson starting with the *pseudo-conceptual* cocktail party talks.

Professor: ...Beautiful weather we are having...it's not raining...in fact it's not snowing, either.
Pupil: That would be rather surprising in the summer.
Professor: Forgive me, Mademoiselle, I was just going to say that...
Pupil: Snow falls in the winter. Winter is one of the four seasons. The other three are...

The weather is of course a universal topic of conversation. Usually, there is no intellectual involvement in conversations about weather or in some mathematical class discussions mentioned above. The *pseudo-conceptual* behavior started immediately after the girl said "winter." This word led her thoughts to the four seasons. However, this association is not relevant at all to the conversation. Mentioning it was an attempt (which eventually failed) to impress the teacher: "I know that winter is one of the four seasons. The other three are...." It turned out that she forgot.

Pupil: I am thirsty for knowledge. And then my parents too, they believe that nowadays a little general culture, however soundly taught, just does not go far enough.
Professor: Modern life has become very complex.
Pupil: And so complicated...

Again, "complicated" as a response to "complex" does not contribute anything to what the professor has just said. If the pupil wanted to express the fact she agreed with what the professor said, a possible way of doing it would be, for instance, by saying: "Yes, indeed." However, the expression "and so complicated" makes the impression that something is added to the statement of the professor. We all know many similar events that happen every day in mathematics class rooms. Mathematics as a special topic is also treated in this lesson.

Professor: ...What is left when you take three from four?
Pupil: Three from four? ...Three from four?
Professor: Yes. That's it. I mean to say what is four minus three?
Pupil: That makes...Seven?
Professor: ...You are muddling it up. Three plus four makes seven, take three away from four and that makes? ...It's not a question of adding up, now you have to subtract.
Pupil: Four.
Professor: No, Mademoiselle. That's not the answer.
Pupil: three then?
Professor: That's not right either, Mademoiselle... I really do beg your pardon....It doesn't make three....I am terribly sorry...
Pupil: Four minus three...three away from four...four minus three? I suppose it wouldn't make ten?

At this situation, it is already clear from the beginning that the girl has no idea about subtraction. In spite of that she tries various strategies which might lead her to the answer. First she tries addition because it was the last operation she used successfully.

This can be considered a *regression*, a common technical term in psychology. Later on, when the addition strategy fails, she tries the guessing strategy by choosing the numbers given in the exercise as answers. Finally, when this strategy fails as well, she chooses a wild strategy. This time she is almost sure it could not bring her to the right answer but she chooses to try it anyway. The show must go on, doesn't it? Now the Professor has to teach the pupil something which she does not know. Let us see how he does it.

Professor: You know how to count all right? You can count up till what number?
Pupil: I can count up to...infinity.
Professor: That's impossible.
Pupil: ...Up to sixteen then.
Professor: Listen. Here are three matches. And here is another one. That makes four. Now watch carefully. There are four of them. I take one away, how many are left?

(The playwright instructions: Neither the matches, nor any of the objects in question are visible. This means that the pupil has to construct all this in her mind.)

Pupil: Five. If three and one make four, four and one make five.
Professor: No....You have a constant predilection for adding up. But it is also necessary to subtract. Integration alone is not enough. Disintegration is essential too. That's what life is. And philosophy. That's science, progress, civilization....All right. We'll take some simpler examples. If you had two noses and I'd plucked one off, how many would you have left?
Pupil: None.
Professor: What do you mean, none?
Pupil: Well, it's just because you haven't plucked one off that I've still got one now. If you had plucked it off, it wouldn't be there anymore.

It is interesting to see that whenever she can, the girl tries to suggest answers which are acceptable by the teacher. If infinity, for instance, is not acceptable she tries sixteen. On the other hand, there are some mental processes of which she is incapable. She can probably not construct in her mind a picture of her with two noses. Hence, she responds to a situation she can imagine: somebody plucking off her only nose. Here, in a way, her reactions stop being *pseudo-conceptual*. She is in a cognitive mode. (The difficulty of assuming counter-reality claims will be discussed further in Chapter 14).

In the following excerpt, we see an interesting development:

Professor: ...Suppose you had only one ear.
Pupil: Yes, and then?
Professor: I stick on another one, how many would you have?
Pupil: Two.
Professor: Good. I stick yet another one on. How many would you have?
Pupil: Three ears.
Professor: I take one of them away, how many ears do you have left?
Pupil: Two.

Professor: Good. I take another one away. How many do you have left?
Pupil: Two.
Professor: No. You have two ears. I take away one. I nibble one off. How many do you have left?
Pupil: Two.
Professor: I nibble one of them off. One of them...
Pupil: Two.
Professor: One.
Pupil: Two.
Professor: One!
Pupil: Two!
Professor: One!!
Pupil: Two!!
Professor: One!!
Pupil: Two!!

Now, the girl does not want or is not able to cooperate with her professor on these issues. The two parties become hostile and there is a first indication of aggression on the side of the professor (I nibble one off). Again, as mathematics teachers, we know quite well this hostility and this tendency to aggression, which we are supposed to suppress. However, the mathematics-education system does not give up. It speaks about "mathematics for all" and it recommends, for instance, that every student will be able to carry out quite sophisticated tasks (NCTM, 1989). The professor does not give up either. Here is another attempt:

Professor: You have...er...you have...er...
Pupil: Ten fingers!
Professor: Good! If you like. Fine! You have ten fingers, then.
Pupil: Yes, Sir.
Professor: How many would you have if you had five of them?
Pupil: Ten, sir.
Professor: No, that's wrong.
Pupil: But I should, Sir.
Professor: I tell you you're wrong!
Pupil: But you have just told me that I have ten...
Professor: And then I told you immediately afterwards that you had five!
Pupil: But I haven't got five, I've got ten!

A similar conflict in a classroom is usually resolved in a civilized manner. In most of the cases, the student will adjust somehow to the educational system. Usually, it is done by adopting a *pseudo-conceptual* behavior (just recall the nine-year-old boy in a previous section of this chapter, who confused addition and subtraction while solving some exercises on the computer). However, this is killing the student intellectually. In Ionesco's lesson there is a real killing of the pupil by the professor

A Meaningful Way to Solve Problems

There is the classic book by Pólya (1945) titled *How to Solve It*. It contains recommendations of how to attack mathematical problems together with many examples. Potentially, it can be a guide book for professional mathematicians as well as for mathematics students.

Unfortunately, as shown in Schoenfeld (1985), very often neither professional mathematicians nor mathematics students follow Pólya's guidance. It happens because the solution process is spontaneous, intuitive, without the inhibition which is required in order to consult Pólya's guidance. Hence, again, my readers are referred to the subtitle of the book—from intuition to inhibition.

Chapter 6
The Pseudo-conceptual Behavior in Contexts Outside the Mathematics Classroom

Carnap's Suggestion to Eliminate Metaphysics and Watzlawick et al. Suggestion to Solve Problems in Human Psychology

Mathematics and science, I believe, are the only contexts where I can diagnose *pseudo-conceptual* and *pseudo-analytical* modes of thinking. This is because we have a relatively clear and well-structured framework to which we can relate when making our cognitive analysis. I am sure that it is much harder, if not impossible, in the cases of literature, philosophy, cognitive psychology, or psychotherapy. As a matter of fact, there was an attempt in philosophy to eliminate what, in my terminology, could be considered *pseudo-conceptual* modes in philosophical thinking, namely, metaphysics. I am speaking about logical positivism and analytical philosophy. The word *analytical* there has, so I believe, a similar meaning to the one I am trying to associate with it in my book. The crucial question, however, is according to what criteria should the analysis be made? It was illustrated in a magnificent way in a classic paper by Carnap (1932): *The Elimination of Metaphysics through the Logical Analysis of Language*. Although many people were convinced by that paper that certain chapters in philosophy should be eliminated, there was no chance to overcome the type of philosophy the logical positivism was fighting. I would like to call this type vague, foggy, or unclear. By doing this, I am emphasizing the subjective aspect of the problem. It turns out that something which is vague, fuzzy, foggy, or unclear to me may be quite clear to somebody else.

And if there is no objective criterion to determine what is clear or unclear, what is conceptual or *pseudo-conceptual* and so on, we will find ourselves again and again in situations where one will claim that a certain text is meaningless while somebody else will claim that it is not only meaningful, but it is also enlightening.

From what I have said till now it is quite clear where I stand in this controversy, but as I said, I cannot think of any argument that will convince somebody to think that a given statement is meaningless while he or she believes that it is meaningful.

S. Vinner, *Mathematics, Education, and Other Endangered Species*,
Mathematics in Mind, https://doi.org/10.1007/978-3-319-90035-3_6

However, it is easy to find such arguments in mathematical contexts. I did it earlier in mathematics classroom situations. The impression was that students use the *pseudo-conceptual* behavior as a survival strategy. They do it in order to survive the process of institutional learning.

Nevertheless, what I call the *pseudo-conceptual* mode of thinking might be for some people a *cognitive style*. As a cognitive style, it exists in all domains and in all contexts, not only in mathematics classes. Here is an example taken from a psychology book called *Change: Principles of Problem Formulation and Problem Resolution* by Watzlawick, Weakland, and Fisch (1974). The text discusses problems in human psychology. The idea is that a psychological problem should be solved by a certain change. The change can be within the problem framework (like improving certain elements in your marriage) or by changing the entire framework (like, in an extreme case, breaking your marriage). This is nice, simple, clever and convincing. There are many illuminating examples later on in their book. However, is this academically impressive? Are there universal rules by means of which a great academic impression can be obtained? One such universal rule is *complexity*. If something can be easily understood, then perhaps it might not be so impressive. And how can such complexity be achieved? One way of doing it is using mathematics. The authors of *Change* "feel that two abstract and general theories, drawn from the field of mathematical logic, may be utilized to help present and clarify some of the conclusions at which they have arrived. These are the *theory of groups* and the *theory of logical types*" (p. 2). The authors declare that they are fully aware that their use of these theories "is far from satisfying mathematical rigor. It should be taken as an attempt at exemplification through analogy" (p. 2). But the problem is not mathematical rigor. The problem is whether the ideas the authors have about group theory or about the theory of logical types are consistent with mathematicians' ideas about these theories.

First, group theory is not a branch of mathematical logic. But this is really unimportant.

Second, it is claimed there (p. 3) that the actual nature of the group members is "irrelevant for the purposes of the theory." The word *purposes* confuses me because I do not understand what it means in this context. I believe that the phrase "irrelevant for the theory" is more appropriate. Usually, a purpose is not associated with mathematical theories.

Later on, after discussing the nature of the group members, the authors discuss the group operation. For some reason, they call it combination. This fact makes me think that somehow we are not talking about the same thing. This impression is strengthened by the following statement: "We can also see that the term *combination* refers to a change from one possible internal state of the group to another" (p. 4). Again, at first sight it looks all right. However, if you look at it again you realize that it claims that applying the group operation to any two of its members changes the internal state of the group. The expression *internal state* is taken from another mathematical context, but it is meaningless in the context of mathematical groups. Applying a notion from one context, in which it is meaningful, to another context, in which it is meaningless, is typical for *pseudo-conceptual* thought processes. Another cause of doing that can be a misconception—a meaningful but wrong idea.

Another claim that is made there (p. 4): "The grouping of things is the most basic and necessary element of our perception and conception of reality. . . .The ordering of the world into groups composed of members who all share an important element in common gives structure to what would otherwise be a phantasmagoric chaos." Probably, nobody would like to argue about the above cognitive claim. However, there does not exist any ground for the attempt to tie grouping of things in the human mind to mathematical groups. This attempt is a result of the fact that a given word (group) is used in a technical context in mathematics. Its everyday meaning is projected on the technical meaning and vice versa. It is quite typical of the way misconceptions are formed in mathematics and science.

All the above continues when the authors speak about the *identity* member (p. 5): "A group contains an *identity* member such that its combination with any other member gives [as a result] that other member[.] which [This] means that it [the identity member] maintains the other member's identity."

Here, the group operation is conceived by the authors as an action upon the group members, an action that changes the members. (Earlier it was claimed that it changes the internal state of the group.) If you act with the identity member, then the identity of the other member is maintained.

As to the theory of logical types the authors point correctly at the "distinction between a member and a class and the fact that a class cannot be a member of itself" (p. 7). On the surface it might look relevant to the statement that "in totalitarian ideologies the individual is seen only as a member of a class and thus becomes totally unimportant," or to the statement that "while the individual members of a species are usually endowed with very specific survival mechanisms, it is well known that the entire species may race headlong toward extinction" (p. 6). Thus although you would like to accept the claims about totalitarian ideologies or about endangered species, you do not need the theory of logical types to be convinced, unless you are in a *pseudo-conceptual* mode of thinking. Namely, you do not understand the concepts involved in the discussion but you are overwhelmed by their use. And why is that? It is because whenever mathematical arguments are involved then they should necessarily be convincing, even when you do not understand the mathematics.

The authors were probably fascinated by the magic of their *pseudo-conceptual* thinking about group theory and the theory of logical types and decided to try to create a similar enthusiasm in the reader. I am sure that some readers were turned on and some of them were turned off by that analogy. Some readers, perhaps, felt impressed by the fact that the authors' analysis is supported or even implied by mathematical theories.

As mentioned above, the authors claim that their use of mathematical theories is "an attempt at exemplification through analogy." But, if it is an analogy, what is the point in presenting it to people whose mathematical background is probably zero?

With this, I conclude my discussion on the use and abuse of mathematics in some social-science contexts.

As for me, the above poor analogy does not disturb my adopting the authors' approach to changes in life as well as in the mathematics educational system. If changes within the system do not achieve our goals, let us change the entire system. Applying this recommendation to mathematics education means that since we have failed in so many cases to teach mathematics in a meaningful way perhaps we should try to change the system. Unfortunately, it is almost impossible to change the system because there are so many people who benefit from the current situation.

Vygotsky's Pseudo-concepts

The word *pseudo* has been used quite often by many authors to denote various things. The common underlying situation is as follows: there is something which is the real thing and there is something else which looks similar but it is quite different from the real thing.

However, the particular meaning of *pseudo* varies from one context to another. The author has to explain both what he means by the real thing and, later on, how the other thing, which looks similar, is different from the real thing. Thus, for instance, Popper (1963), in his *Conjectures and Refutations*, explains what he means by *real science* (a refutable theory) and *pseudoscience* (a theory which is not refutable).

Vygotsky (1986), in his *Thought and Language*, suggested a concept-formation theory. According to his theory, the concept-formation process has stages. "The ascent to concept formation is made in three basic phases, each divided in turn into several stages" (p. 59). This is not the place to present and explain Vygotsky's theory of concept formation. However, the final stage before the "real" concept, the bridge to it, is called "the pseudo-concept" (pp. 66–69). Thus, the notion *pseudo-conceptual* that I am using in this book is basically different from Vygotsky's *pseudo-concepts*. According to Vygotsky the subject is following a "cognitive track." This is not necessarily the case with mathematics students who act in a *pseudo-conceptual* mode of thinking. These students, intentionally or unintentionally, can be miles away from the cognitive track. They might use all kinds of non-conceptual steps in order to come up with an answer to a question posed to them by a teacher or by an outside examiner (like in a matriculation exam). It is quite possible that the *pseudo-conceptual* elements which are used by the students might lead them to a *true-conceptual* mode of thinking; however, it is not necessarily so. That is the main difference between Vygotsky's pseudo-concepts and the *pseudo-conceptual* modes described in this book.

On the other hand, there is some similarity to Vygotsky's pseudo-concepts in the following way: The pseudo-concept looks very often like a real concept. It is hard to distinguish between the two. Vygotsky deals with concepts in which grouping and classification are involved (specifically, concepts related to height and size, like "tall-and-large," and grouping objects by these concepts). For those, he even designed an experiment, the results of which he reported in the chapter "An experimental Study of the development of Concepts" of his above book.

About the difficulty to distinguish between the pseudo-concept and the concept in that experiment there is a note, the essence of which is that only some crucial questions can reveal the true nature of the subject's thought. "The lack of a true conceptual foundation is not revealed until the subject is required to put in operation the ideas that underlie this grouping" (p. 66).

When discussing pseudo-concepts versus concepts, Vygotsky claims that "although the results [with pseudo-concepts] are identical, the process by which they are reached is not at all the same as in conceptual thinking" (p. 66).

In this sense, my claims about *pseudo-conceptual* processes and *true-conceptual* processes are identical with Vygotsky's. *Conceptual* behaviors and *pseudo-conceptual* behaviors may look the same, even though the thought processes that produce them are different. However, as we saw in chapter 5, by posing additional questions it is possible to identify the difference between *true-conceptual* and *pseudo-conceptual* thought processes.

The word *pseudo* has a negative connotation in everyday language. It is not supposed to have such connotation in the above technical contexts. It is hard to avoid it, though, because of the impact of everyday language. The reader should try to overcome this spontaneous tendency to consider *pseudo* as negative. It might be negative in the case of somebody who tries intentionally to use *pseudo-conceptual* thought processes in order to get some kind of credit. However, if *pseudo-conceptual* thought processes are used in a spontaneous way with no intention of pretense, there is nothing negative about it.

Chapter 7
Can E-learning (Computer-based Learning) Replace Teacher–Student Individual Communication?

It is quite clear that teachers must know the mathematics they teach. However, this is not enough. They should also understand the way students think. They should be aware of potential mistakes and misconceptions and, by means of discussions with their students, prevent them from making these mistakes and from forming undesirable misconceptions. Therefore, e-learning cannot replace the human interaction between teachers and students.

Erlwanger's paper about Benny (1973) is convincing evidence of that. In 1973, e-learning did not exist as it exists today. However, we can consider the teaching strategy of the program in which Benny was involved a kind of e-learning. The title of this program was: *Individually Prescribed Instruction (IPI) Mathematics*.

In this kind of instruction, the teacher, as a matter of fact, became an administrator. He or she gives the pupils class assignments. The pupils are supposed to solve them. If they solve about 70% of the exercises correctly, the teacher gives them the next level of assignment. If not—the pupils get more exercises at the same level.

Undoubtedly, nowadays the learning software is much more sophisticated. However, my claim is that the teacher should be involved in the learning procedure in order to prevent the students from making the mistakes of the kind which were made by Benny.

Benny's task was to convert simple fractions to decimals and vice versa. Here are some excerpts from a conversation between Erlwanger (denoted by E hereafter) and Benny (denoted by B hereafter):

E: How would you write $\frac{2}{10}$ as a decimal or a decimal fraction?
B: One point two (writes 1.2).
E: And $\frac{5}{10}$?
B: 1.5.

(Benny's method is to add the nominator and the denominator and put a decimal point somewhere.)

S. Vinner, *Mathematics, Education, and Other Endangered Species*, Mathematics in Mind, https://doi.org/10.1007/978-3-319-90035-3_7

At this point, Erlwanger claims that "this method enabled Benny to convert any fraction to a decimal. Some of the answers he gave were: $\frac{429}{100} = 5.29$, $\frac{3}{1000} = 1.003$, $\frac{27}{15} = 4.2$, $\frac{1}{8} = 0.9$, $\frac{1}{9} = 1.0$, and $\frac{4}{6} = 1.0$. Benny applied this method consistently. Moreover, he was fully aware of the fact that he got equivalent results for fractions of different values, but he did not appear to think that there was anything wrong with that." This is illustrated in the following excerpt:

E: and $\frac{4}{11}$?
B: 1.5.
E: Now does it matter if we change this $[\frac{4}{11}]$ and say that is eleven fourths? [E writes $\frac{11}{4}$].
B: It won't change at all; it will be the same thing…1.5.
E: How does it work? $\frac{4}{11}$ is the same as $\frac{11}{4}$?
B: Ya…because there is ten at the top. So you have to drop that 10…take away the 10; put it down at the bottom. [Shows $\frac{11}{4}$ becomes $\frac{1}{14}$]. Then there will be 1 and 4. So, really it will be $\frac{1}{14}$. So you have to add this numbers up which will be 5; then 10…so 1.5.

At this point, I would like to make a comment about a communication failure between Erlwanger and Benny. While Erlwanger was trying to form a *cognitive conflict* (a notion suggested by Piaget), which was supposed to cause Benny to rethink his algorithm and to realize that there is something wrong with it, Benny was busy trying to explain to Erlwanger his algorithm.

Here are more excerpts from Erlwanger's paper: "When Benny failed to get 80% marked right by the IPI aide, he tried to grasp the pattern of the correct answers; he then quickly changed his answers in ways which he hoped would better agree with the key.

"Benny's case indicates that a 'mastery of content and skill' does not imply understanding. This suggests that an emphasis on instructional objectives and assessment procedures alone may not guarantee an appropriate learning experience for some pupils." An additional dialogue excerpt:

E: Finding answers seems to be like a game.
B: (Emotionally) Yes! It's like a wild goose chase.
E: So you're chasing answers that the teacher wants?
B: Ya, Ya.
E: Which answers would you like to put down?
B: [shouting] Any! As long as I knew it would be the right answer. You see. I am used to checking my own work and I am used to the key. So I just put down $\frac{1}{2}$ because I don't want to get it wrong.
E: Mm…
B: Because if I put $\frac{1}{4} + \frac{1}{4}$, they will mark it wrong. But it would be right. You agree with me there, o.k. If I put $\frac{2}{4}$, you agree there. If I put $\frac{1}{2}$, you agree there too. They are all right….They mark it wrong because they just go by the key. They don't

go by if the answer is true or not. They go by the key. It'd be like if I had $\frac{2}{4}$. They wanted to know what it was, and I wrote down one whole number, and the key said a whole number, it would be right; no matter that it was wrong.

(Note that Benny becomes quite emotional at this part of the conversation. It seems that he already developed a negative attitude to mathematics and to the way it is taught. Well, as I have mentioned several times in this book, for the majority of adults, learning mathematics is conceived as a negative experience.)

Furthermore: "Benny's views about rules and answers reveal how he learns mathematics. Mathematics consists of different rules for different types of problems. These rules have all been invented. But they work like magic because the answers one gets from applying these rules can be expressed in different ways, 'which we think are different but really they're the same'. Therefore, mathematics is not a rational and logical subject in which one has to reason, analyze, seek relationships, make generalizations, and verify answers. His purpose in learning mathematics is to discover the rules and to use them to solve the problems. There is only one rule for each type of problem, and he does not consider the possibility that there could be other ways of solving the same problem. Since the rules have already been invented changing a rule was wrong because the answer 'would come out different.'"

A consequence of Erlwanger's study is that "Benny's misconceptions indicate that the weakness of IPI stems from its *behaviorist* approach to mathematics, its instruction, and its concept of individualization."

(It is worthwhile to inform the readers that during the 60s of the twentieth century *behaviorism* was the dominant paradigm. Social scientists and psychologists tried to avoid the four letter dirty word *mind*. Like natural scientists, they tried to speak about outcomes. Their research schema was: observable stimulus, observable response. Their teaching paradigm was: stimulus, response, and reward.

This paradigm was extremely successful with animals that appear in circuses. The idea of treating human beings as animals is not accepted nowadays. However, in the 60s of the last century it was quite prevalent in the community of psychologists. One of them, Skinner (1904–1990), even tried to educate his daughter according to this paradigm. Thus, Erlwanger's paper was a major breakthrough in educational research. It can be considered as a revolutionary paper which helped the educational research get rid of the *behaviorist* paradigm in education.)

Erlwanger's claim about the views Benny had on mathematics is quite relevant to Skemp's distinction between instrumental understanding and relational understanding (Skemp, 1976, 1979). In an equivalent terminology, it is the same distinction between procedural versus conceptual understanding.

Many students care about instrumental understanding (that is—knowing *how*), but they do not care at all about relational understanding (that is—knowing *why*). The attempts of Benny to find answers that satisfy "the key" can be considered with what I called the *pseudo-conceptual* behavior. The students are looking for verbal cues that will lead them to an answer that will please the teacher (see the section titled "Verbal Cues" in Chapter 5).

An additional comment which is worthwhile to mention here is the difficulty of convincing students to abandon their mistaken way of thinking. Rosnick and Clement (1980) reported in their paper about the treatment that they gave to the students who made mistakes in the "Students and Professors" problem ("Write an equation using the variables S and P to represent the following statement: 'There are 6 times as many students as professors at this university'. Use S for the number of students and P for the number of professors.")

The common mistake was: $P = 6S$.

Rosnick and Clement gave the students a treatment which included manipulatives and other conceptual explanations.

Here is an excerpt from an interview from Rosnick and Clement's paper:

Dawn (intervewed after the treatment, solving the Oil and Vinegar problem where there is three times as much oil as vinegar in a salad dressing): My first impulse would be to write three times: three times O equals V. . . .So then, because that's wrong, I would change it to 3V = O because I know it's the other way around. . .

Teaching is tough when you have students who are not interested in meaningful learning but only in passing exams.

Thus, the bottom line of Erlwanger's paper, as mentioned above, is: "Benny's misconceptions indicate that the weakness of IPI stems from its *behaviorist* approach to mathematics, its instruction, and its concept of individualization."

This is also true about e-learning, which is a computerized version of the IPI program.

Chapter 8
What Is Mathematics?

The question may sound quite embarrassing to some people. Isn't it like asking what music is, what poetry is or what art is? However, there is a clear answer to the question of what mathematics is.

Mathematics is a collection of mathematical theories. Each theory is a set of theorems and has models in which the theorems hold. There is also a deductive system for proving theorems.

Among these theories, one can mention the theory of groups, the theory of rings, the theory of fields, the theory of probability, and so on.

Now I will explain in more detail what a mathematical theory is, what models are, and what a deductive system is.

Well, a model consists of a set of abstract elements, operations on these elements, and relations between these elements. The theorems of the theory are claims about the elements of that set, concerning the relations and the operations on the set. The elements, the relations, and the operations of the model satisfy the theorems of the theory.

Every theorem is either a primary theorem, called axiom, or a provable theorem. The theorems are provable by means of axioms, of previously proved theorems, and of the rules of logic. This is a deductive system. (In fact, the deductive system also includes, what I called in previous chapters, a deductive structure. I am not going to elaborate on it here.)

This sounds quite scary but as a matter of fact all of us know such theories and their models from our studies in elementary school and in high school.

The simplest theory is school arithmetic, which everybody becomes familiar with by the end of the sixth grade or even before. Its model is the set of natural numbers (including zero), 0, 1, 2, 3, ...The operations are addition, subtraction, multiplication, and division, and the relations are the equality relation and the inequality relation (i.e, a certain number is less than another number, usually denoted by $a < b$). The basic laws (the axioms) of this theory are as follows:

1. $a + b = b + a$ (commutativity of addition),
2. $ab = ba$ (commutativity of multiplication),

S. Vinner, *Mathematics, Education, and Other Endangered Species*, Mathematics in Mind, https://doi.org/10.1007/978-3-319-90035-3_8

3. $(a + b) + c = a + (b + c)$ (associativity of addition),
4. $(ab)c = a(bc)$ (associativity of multiplication),
5. $(a + b)c = ac + bc$ (distributivity of multiplication over addition),
6. $a + 0 = a$, and
7. $1 \cdot a = a$.

Every sixth grader is familiar with these laws. To these laws is added sometimes in the 12th grade the axiom of mathematical induction. This axiom causes many students a lot of troubles and I wonder about the wisdom behind teaching it. Only students who study in an accelerated track can cope with it. However, despite the possible difficulties, I would like to present it here. But before doing it, I would like to give an example of what can be done with laws (axioms) 1–7 above. In the tenth grade, students usually study three multiplication formulas (theorems) which are:

1. $(a + b)^2 = a^2 + 2ab + b^2$,
2. $(a - b)^2 = a^2 - 2ab + b^2$, and
3. $(a + b)(a - b) = a^2 - b^2$.

While teaching mathematics it is emphasized that mathematical claims should be proved unless the proofs are too complicated. In case they are—the student is asked only to remember the claim. The case of the multiplication formulas is quite simple. Here is the proof of the first one:

$$(a + b)^2 = (a + b)(a + b).$$

By means of the distributive law, we get:

$$(a + b)(a + b) = (a + b)a + (a + b)b.$$

Using the distributive law again, we get:

$$(a + b)a + (a + b)b = aa + ba + ab + bb.$$

By the commutative law of multiplication ($ba = ab$), we get:

$$aa + ab + ab + bb = a^2 + 2ab + b^2.$$

The other two multiplication formulas are proved in a similar way.

Thus, as I claimed, many arithmetical theorems can be proved by 1–7 above. Unfortunately, in order to prove many other theorems the axiom of mathematical induction is required. So here, comes the "monster."

First, I will formulate it and then I will explain why it is a monster for many students. The axiom of mathematical induction:

Let P be any property of natural numbers. If the number 1 has the property P, and if it can be proved that if a natural number n has the property P, then the number $n + 1$ has it too—then all the natural numbers have the property.

At first sight, it looks scary because it is very long. In order to digest it, one has to cut it down to several parts:

The axiom of mathematical induction infers that if we complete a proof with the following parts A and B, then a property P of a natural numbers n holds for all natural numbers. Steps A and B of the proof are:

A. Prove that the property P holds for the first natural number $n = 1$.
B. Prove that if the property P holds for some natural number n, then the property also holds for $n + 1$.

Another difficulty is the abstract property P. Let us work on this. We all know some properties of natural numbers. For instance: Being even, being odd, being prime (2, 3, 5, 7, . . .), being a square (1, 4, 9, 16, 25, . . .), and many more. But there are many properties of natural numbers which do not look like that. Properties that we will deal with are equalities with n. The fact that an equality holds for n can be considered a property of n.

Now I would like to remind the reader of the definition of an arithmetic progression, already mentioned in Chapter 1. An arithmetic progression is a sequence such that each element beside the first one is obtained from its predecessor by adding a constant number. We denote the nth element of (any) sequence by a_n and the constant (which is the difference between two consecutive elements) by d.

Using this notation, the equality $a_{n+1} = a_n + d$ expresses how the $(n + 1)$th element can be calculated in terms of its predecessor.

A claim: The nth element of an arithmetic progression can be calculated by the formula: $a_n = a_1 + (n - 1)d$. This formula will be the property P in our proof by the axiom of mathematical induction:

A. When $n = 1$ the formula becomes $a_1 = a_1 + (1 - 1)d$, hence the property is true for $n = 1$.
B. Assume the property is true for n, namely, $a_n = a_1 + (n - 1)d$. Using this assumption, we will prove that it is also true for $n + 1$. Namely, that $a_{n+1} = a_1 + ((n + 1) - 1)d$, or that $a_{n+1} = a_1 + nd$.

   Using the assumption $a_n = a_1 + (n - 1)d$ to substitute for a_n in $a_{n+1} = a_n + d$ we get: $a_{n+1} = (a_1 + (n - 1)d) + d = a_1 + nd$

   So, we have completed step B.

It follows from the axiom of mathematical induction that the formula (property) is true for all natural numbers n.

Q.E.D (which means we have proved what we had to prove).

Note that in order to prove the formula for the nth element of the arithmetic progression we did not have to use the axiom of mathematical induction. The fact that the nth element of the arithmetic progression is achieved by adding $n - 1$ times the constant difference d is a direct consequence of the definition of arithmetic progression. However, I used mathematical induction in order to practice it as preparation for more complicated cases.

The first one of them will be the sum of the first n elements in an arithmetic progression.

Theorem: If $a_1, a_2, a_3, \ldots, a_n$ are the first n elements in an arithmetic progression then for all natural numbers n:

$$a_1 + a_2 + \ldots + a_n = \frac{n(2a_1 + (n-1)d)}{2}.$$

Proof by mathematical induction:

A. When $n = 1$ the sum on the left-hand reduces to a_1.
On the other hand, if we substitute $n = 1$ in the right-hand side of the formula, we get: $a_1 = \frac{1(2a_1+(1-1)d)}{2} = \frac{2a_1}{2} = a_1$. Hence, the claim is true for $n = 1$.

B. Assume it is true for n, namely:

$$\text{(I)} \quad a_1 + a_2 + \ldots + a_n = \frac{n(2a_1 + (n-1)d)}{2}.$$

Using this assumption, we will prove it for $n + 1$. Namely, we have to prove that: $a_1 + a_2 + \ldots + a_n + a_{n+1} = \frac{(n+1)(2a_1+((n+1)-1)d)}{2}$, or that:

$$\text{(II)} \quad a_1 + a_2 + \ldots + a_n + a_{n+1} = \frac{(n+1)(2a_1 + nd)}{2}.$$

Using assumption (I), we can replace $a_1 + a_2 + \ldots + a_n$ in (II) by $\frac{n(2a_1+(n-1)d)}{2}$, and rewrite (II) as:

$$\frac{n(2a_1 + (n-1)d)}{2} + a_{n+1} = \frac{(n+1)(2a_1 + nd)}{2}.$$

We know that $a_{n+1} = a_1 + nd$, thus what we have to prove now is that:

$$\frac{n(2a_1 + (n-1)d)}{2} + a_1 + nd = \frac{(n+1)(2a_1 + nd)}{2}.$$

Using our algebraic skills to remove the parentheses, we will find out that the results on both sides is: $\frac{2na_1+n^2d+2a_1+nd}{2}$. This completes part B.

Thus, we have proved what we had to prove.

Q.E.D.

Now I am going to introduce the most complicated case in this chapter: The formula for the sum of the first n squares (for all natural numbers n). It is

$$1^2 + 2^2 + 3^2 + \ldots + n^2 = \frac{n(n+1)(2n+1)}{6}.$$

This formula will be the property in our proof by the axiom of mathematical induction.

Proof:

A. For $n = 1$, the statement reduces to $1^2 = \frac{(1 \cdot 2 \cdot 3)}{6}$, which is true. Hence, $n = 1$ has this property.
B. Let us assume that the number n has it. Namely, we assume that the following is true:

$$\text{(I)} \quad 1^2 + 2^2 + 3^2 + \ldots + n^2 = \frac{n(n+1)(2n+1)}{6}.$$

We have to prove that this assumption implies that it is also true for $n + 1$. Namely, we should prove that the following is also true (we got this equation by substituting $n + 1$ for n in (I)):

$$\text{(II)} \quad 1^2 + 2^2 + 3^2 + \ldots + n^2 + (n+1)^2 = \frac{(n+1)(n+2)(2n+3)}{6}$$

So, by assumption (I) we can replace $1^2 + 2^2 + 3^2 + \ldots + n^2$ in (II) and get the equivalent equality:

$$\text{(III)} \quad \frac{n(n+1)(2n+1)}{6} + (n+1)^2 = \frac{(n+1)(n+2)(2n+3)}{6}.$$

To prove it, let us use our algebraic skills for adding the algebraic fractions on the left-hand side (6 is the common denominator):

$$\frac{n(n+1)(2n+1)}{6} + (n+1)^2 = \frac{n(n+1)(2n+1) + 6(n+1)^2}{6}.$$

By removing parentheses, we get after several steps:

$$= \frac{2n^3 + 2n^2 + n^2 + n + 6n^2 + 12n + 6}{6} = \frac{2n^3 + 9n^2 + 13n + 6}{6}.$$

Now, is $\frac{2n^3+9n^2+13+6}{6}$ equal to $\frac{(n+1)(n+2)(2n+3)}{6}$ (the right-hand side of (II))? Let us find out by removing the parentheses on the right-hand side. So:

$$\frac{(n+1)(n+2)(2n+3)}{6} = \frac{(n^2+3n+2)(2n+3)}{6}$$
$$= \frac{2n^3 + 6n^2 + 4n + 3n^2 + 9n + 6}{6} = \frac{2n^3 + 9n^2 + 13n + 6}{6}.$$

And this is exactly what we got for the left-hand side of (II).

We have completed part B of the proof.

Thus, the formula is true for every natural number.

Q.E.D.

The reader can see that the above proof is much longer than the common proofs the students have been used to until that stage of their studies. Moreover, although I tried to justify the majority of the steps in the proof I have avoided mentioning all of them because I did not want the proof to be too clumsy.

I hope that by now I have explained to the reader what mathematics is. The question is what does the majority of people think about mathematics. I hardly remember what I thought about mathematics when I was a high-school student. I had a wonderful mathematics teacher who was for me a role model when I started to teach mathematics and for many years later. But he never got into the question of what mathematics was. Reconsidering it after many years, I probably thought that mathematics was a collection of procedures supposed to be used for homework assignments and for all kinds of mathematics tests, especially the matriculation exams.

In Chapter 13, I will discuss what elementary mathematics teachers think about the question of what mathematics is.

Summing up, mathematics is a collection of theories about different domains, as I described above. Among these theories are group theory, ring theory, and field theory. Usually, teachers do not use this terminology, but, even so, high-school students know examples of groups, rings, and fields (namely, models of these theories).

Groups that high-school students know: (1) the set of integers with addition, (2) the set of rational numbers with addition, (3) the set of real numbers with addition, (4) the set of rational numbers excluding zero with multiplication, and (5) the set of real numbers excluding zero with multiplication.

Rings that high-school students know: (1) the set of integers with addition and multiplication, (2) the set of rational numbers with addition and multiplication, and (3) the set of real numbers with the same operations.

Fields that high-school students know: (1) the set of rational numbers with addition, multiplication, and division with one exception—division by zero is not allowed, and (2) the set of real numbers with the same operations and same exception.

It is quite clear to me why mathematics teachers avoid mentioning these concepts to their students. Knowing these concepts will not help the students in doing their homework assignments or passing the examinations which they have to go through. However, it prevents their students from getting a wider view about what mathematics really is.

A final comment for this chapter: It is possible to explain what mathematics is by pointing at some examples. This was done in Courant and Robbins (1941). Although the various chapters of their book presented to the readers various mathematical theories, it probably does not enable the reader to make the generalization and arrive at the definition that I have given: What is mathematics? My answer was: It is a collection of theories.

Chapter 9
The Role of Examples in the Learning of Mathematics and in Everyday Thought Processes

This chapter can be classified as an introspective chapter. It is written, like many other chapters in this book, for teacher trainers and for mathematics teachers, as well as for readers interested in understanding how concepts and conjectures are formed in our minds. They are invited to reflect with me on the issues presented in the chapter. The results of their reflections do not have to coincide with mine.

Consider, for instance, two people discussing the weather. The concepts that are involved can be temperature, rain, clouds, etc. The conjecture can be: it will not rain within the next half an hour. The arguments that can be brought to support this conjecture can be: There are no clouds in the sky; there is no wind; the temperature is quite high, etc. I am not claiming that these arguments are good enough to establish the conjecture; these are only examples. In short, in thought processes that are similar to the above, there are two participants: the concepts and the conjectures about them. Hence, in a cognitive analysis of such thought processes, we are supposed to ask: (1) How are concepts formed in our minds? (2) How are conjectures formed in our minds? and also, (3) How do conjectures obtain their validity in our minds? The point that I would like to make in this chapter is that examples have a crucial role in the cognitive processes associated with these three questions.

I shall start dealing with each of these processes in everyday thought processes, and then I will switch to mathematical thought processes. At the end of the chapter, I will discuss everyday thought processes in the light of mathematical thought processes.

S. Vinner, *Mathematics, Education, and Other Endangered Species*, Mathematics in Mind, https://doi.org/10.1007/978-3-319-90035-3_9

Concept Formation by Means of Examples in Early Childhood

Concept formation is one of the main issues in cognitive psychology and philosophy. However, I would like to avoid the sophistication typical of psychological or philosophical papers. My main concern here is the relation between the concept and its name. The concept is an idea evoked in our minds when we hear or see the concept name. The concept name is a linguistic entity, usually called a notion or a term. It has a physical aspect, visual or vocal. The concept, on the other hand, exists only in our minds. The words *concept* and *notion* are often used interchangeably, with the context resolving the ambiguity. I am restricting myself here to special cases in which concepts are formed by means of examples. These cases can be characterized by the following: In order to form the concept in our minds, we are supposed to construct a class of objects (concrete or abstract) or a class of occurrences of the concept to which the concept name applies.

Take for instance little children (starting at the age of few months) who are about to learn the names of basic concepts in their daily experience. The common practice of teaching children the concept of a chair, for instance, is to point at various chairs in various contexts and to say "chair."

Amazingly enough, after some repetitions, the children understand that the word *chair* is supposed to be related to chairs, which occur to them in their daily experience, and when being asked "what is this?" they understand that they are expected to say: "chair." Later on, they will imitate the entire ritual on their own initiative. They will point at chairs and say: "chair." I would like to make a theoretical claim here by saying that they have constructed in their minds the class of all possible chairs. Namely, a concept has been formed in their minds, and whenever a concrete object is presented to them, they will be able to decide whether it is a chair or not. Of course, some mistakes can occur in this concept-formation process. This is because in this process two cognitive mechanisms are involved. The first mechanism is the one that identifies similarities. The mind notices that one particular chair is similar to another particular chair and in each case the word *chair* was used by the adult at the end of the ritual. The second mechanism is the one that distinguishes differences. The mind distinguishes that a certain object is not similar to the chairs which were presented to the child, and therefore the child is not supposed to say "chair" when an object that is not a chair is presented to him or her by the adult.

Mistakes do occur quite often. They occur either because the mechanism that is supposed to identify similarities fails or because the mechanism that is supposed to identify differences fails.

An object that is not a chair (say a small table) appears to the child (or even to an adult) like a chair. In this case, the object will be considered an element of the class of all chairs while, in fact, it is not an element of this class. The second reason for mistakes is that an object that is really a chair will not be identified as a chair because of its weird shape.

Another example of this type is as follows: Sometimes, children mistake dogs for cats and vice versa. These are intelligent mistakes because there are some similarities between dogs and cats. They are both animals; sometimes they have similar size (in the case of small dogs) and so on.

Much more complicated are concepts such as love, beauty, justice, and wisdom.

I would like to call the above process which leads, in our minds, to the construction of the class of all possible objects to which the concept name can be applied, a generalization schema. This schema involves the mechanisms of similarity and differences described above. (I will discuss a similar generalization schema involved in the formation of conjectures in a later section of this chapter.) Thus, generalizations are involved in the formation of any given concept. Therefore, concepts can be considered generalizations.

The actions we use to teach children the concept of chair—pointing and naming—are called *ostensive definitions*. Of course, only a narrow class of concepts can be acquired by means of ostensive definitions. Other concepts are acquired by means of explanations, which can be considered definitions at this stage. Among these concepts I can point, for instance, at a forest, a school, work, hunger, and so on.

When I say definitions at this stage, I do not mean definitions which are similar, or even seemingly similar, to rigorous mathematical definitions. The only restriction on these definitions is that familiar concepts will be used in order to explain a non-familiar concept; otherwise, the explanation is useless. (This restriction, by the way, also holds for mathematical definitions, where new concepts are defined by means of previously defined concepts or by primary concepts.) In definitions which we use in a non-technical context in order to teach concepts we can use examples. For instance, in order to define furniture we can say: "Chairs, beds, tables, desks, couches, etc., are furniture."

The description that was just given deals with the primary stage of concept formation. However, concept formation in ordinary language is far more complicated and very often, contrary to the mathematical language, may end up in a vague notion. Take, for instance again, the notion of furniture. The child, when facing an object which was not previously introduced to him or to her as furniture, should decide whether this object is a piece of furniture or not. He or she may face difficulties doing it. Adults might have similar difficulties. This is only one example out of many which demonstrates the complexity of concept formation in the child's mind as well as in the adult's mind. There are even greater complexities when concept formation of abstract nouns, adjectives, verbs, and adverbs is involved. Nevertheless, despite that complexity, the majority of children acquire language at an impressive level by the age of six (an elementary level is acquired already at the age of three).

However, what I have said about simple concepts such as chair and table is enough for my purpose in this book, which is a mathematics education book. Quine (1964) is an illuminating source for further discussions about concept formation for the interested reader.

In addition to the language acquisition the child also acquires broad knowledge about the world. He or she knows that when it rains it is cloudy, they know that dogs bark and so on and so forth. In short, they know infinitely many facts about their environment. And again, it is obtained in a miraculous way, smoothly without any apparent difficulties. We will see later in this chapter that things become awkward when it gets to mathematics.

Concept Formation by Means of Explanations and Examples

When new concepts are introduced to older children or to adults, sometimes general explanations are given. In order to clarify the meaning of these concepts, some examples are presented. These cases are quite typical of formal learning in all ages. For instance, if a music teacher wants to explain to his or her pupils (even kindergarten children) the notion of ritardando, he or she might say: "It is slowing down the melody," and then the teacher can give the pupils some examples by singing or playing for them a certain melody in which he or she slows down at a certain point. The examples help the children understand the concept, namely, the meaning of the word.

In a poetry class, when teachers want to explain the notion of metaphor, they can say: "Metaphors are comparisons that show how two things that are usually not alike are similar in a way."

They give examples like the following:

1. "Love is a rose."
2. "My mathematics teacher is a pain in the ass."
3. "I am an endangered species."

Again, the examples help us understand the general explanation, which is not sufficient for many students. For many students, the explanation is not useful to distinguish between metaphors and the similar notion similes. Therefore, as in the cases mentioned above some mistakes might occur. Comparisons which are not metaphors can be considered elements of the class of all metaphors and vice versa. Many people consider "love is like a rose" a metaphor, although it should be considered a simile, because the comparison is formed by means of the word *like*. Thus, even if these people were exposed to the characterization of similes in a literature class or saw it in a dictionary, they would forget the relevant explanation or, in case they did not forget it, ignore the small difference: the use of the word *like*, which determines whether a given comparison is a metaphor or a simile. I myself, sometimes, cannot decide whether a given comparison is a simile or an analogy (another type of comparison). So, in many cases, general explanations do not help determine uniquely the class of all occurrences, which is the concept (the meaning of the notion). In addition, we have a whole class of vague notions, namely, notions that do not uniquely determine their concepts. In the case of vague notions, different people may have different views about the validity of applying them to a certain example. This kind of notions will be discussed in the next section.

Cases in Which Explanations of Notions Do Not Help Solve Disagreements About Them

There are notions about which there are serious disagreements on whether they should be applied to certain occurrences of the notion. Take for instance, the notion of a poem. It can be claimed that a poem is "a verbal composition designed to convey experiences, ideas, or emotions in a vivid and imaginative way" (American Heritage® Dictionary of the English Language, Fifth Edition). Some people, usually experts in humanities, even consider this characterization a definition.

Still, there are serious controversies concerning contemporary texts, which are presented to the public by their authors as poetry and strongly rejected as poetry by others. The same holds for certain items that are displayed in modern-art museums. The controversy is whether these items can be considered art.

Another example that is more related to mathematics education, as well as to other contexts in education, is the notion of *metacognition*. Metacognition can be defined as "cognition about cognition" or "knowing about knowing" (Wikipedia, Metacognition, 11 May, 2018).

One of the theoretical questions about metacognition is at what age (if at all) a child is capable of metacognitive thinking. Some researchers claim that even little children are capable of metacognitive thinking and, in order to support their claim, they point at expressions such as: "This was very hard for me to do." Other researchers refuse to accept it as an example of metacognitive thinking.

Notions like those just mentioned belong to a higher level of intellectual discourse. They can even be considered as technical notions although they can appear in everyday conversations of people who went through some formal studies (high school or college).

Concept Images and Concept Explanations in Common Everyday Thought Processes

It can be claimed about all the concepts discussed above that they are formed in our minds in the course of language development by means of examples. At the first stage, we perceived examples and then came the notion which denoted them. In other words, by a social convention that we implicitly accepted, a collection of different examples became a class in our minds. The explanations came later in the course of language development.

In many cases, the explanations were redundant. The concept had been acquired by means of examples. Now, when we hear or see an occurrence of a notion, various associations are evoked in our minds. They might be verbal, visual, or even vocal (additional senses may be involved in some cases). They can be emotional as well. All these associations of the notion were called in an earlier work of mine (Vinner, 1983) *concept images*. My claim is that in common thought processes, the move from one stage of concept formation to the next one is determined by *concept images*. This is especially true about thought processes that are spontaneous, and therefore uncontrolled by analytical thinking.

For instance, assume that somebody asks you: What kind of vehicles do people in Prague use for transportation? In order to answer it you do not consult the explanations of the notion of *vehicle* that you have in your mind. You use your *concept image* of *vehicle* and you answer accordingly. You may say: "buses, taxis, motorcycles, trams." Then, the conversation can be continued with an additional question: "What about underground trains or bicycles?" And, again, you will not consult the explanation of "underground trains." Your answer will be determined by your *concept images*. Thus in common everyday thought processes, *concept definitions* very rarely have any meaningful role, if at all. Although it is trivial, it is worthwhile mentioning that *concept images* are personal. They may vary from one person to another. They may also vary from one culture to another. For instance, in Western cultures (to the best of my knowledge) when you say "cappuccino," it refers to a hot drink. In Taiwan, it refers to either a hot or a cold drink.

Another example of the same kind is as follows: In Western cultures that I know, *plain yogurt* means that nothing is added to it. In Taiwan, it means that only sugar is added to it.

All of the above was related to concepts. Now I would like to say a few words about problem solving in everyday life.

Using our everyday experience, we develop strategies to solve problems we face in everyday situations. Pólya (1945) developed strategies to solve mathematical problems. But each individual should develop his or her own strategies to solve problems they may face in their everyday lives. I will not bother my readers with my strategies.

I just recommend to them to reflect on their own strategies, which in some cases were developed implicitly and in other case were developed explicitly.

Concept Images and Concept Definitions in Mathematical Thought Processes

The role of *concept definition* in mathematical thought processes is dramatically different from the role of concept explanations in everyday common thought processes. We are speaking here about the desirable mathematical thinking, the one that follows from the nature of mathematics as a deductive theory. However, for children who only start studying mathematics at elementary school with thinking habits that they acquired in early age, this is not a simple task.

One crucial difficulty in mathematical thinking is that, as claimed above, mathematical concepts are strictly determined by their definitions.

In the course of their mathematical studies, children, quite often, are presented with mathematical notions with which they were familiar from their past experience.

For instance, in kindergarten they are shown geometrical figures such as squares and rectangles. The adjacent sides of the rectangles which are shown to the children in kindergarten have always different lengths.

In third grade, in many countries, the teacher presents to the children the following definition: A rectangle is a quadrangle that has four right angles. According to this definition, a square is also a rectangle (it depends, of course on the state curriculum). When it happens, a conflict may be formed in the child's mind between the suggested definition and the concept he or she already has about rectangles.

The concept in the child's mind was formed by the set of examples and the properties of these examples which were presented to the child. In fact, that is the child's *concept image*. Thus, there is a cognitive conflict between the *concept image* and the *concept definition*, and it requires a conceptual change.

Unfortunately, quite often, the desired conceptual change does not occur.

It is the task of the third grade teacher (whose mathematical background, sometimes, is not satisfactory) to explain to the children why, from this point on, squares should be considered rectangles too.

Later on, or at the same stage, they are required to treat rectangles as parallelograms, while their *concept images* tell them that parallelograms do not have right angles. These are situations in which children may start to develop an ambivalent attitude toward mathematics. Accepting rules of thought without understanding them may lead to detachment and even hostility. In fact, the detachment and hostility start in many children already at the elementary level.

I am not going to point at more situations at the elementary level in which an ambivalent attitude toward mathematics can be developed. Also, I am not suggesting in this book cures to the problem. My only recommendation is to borrow some advice from the relatively new social-science discipline—conflict management. The advice is that while interacting with concepts that you have a conflict with, if possible, try to focus on issues about which it is relatively easy to agree, and try to avoid, as long as possible, dealing with issues that are extremely hard to solve. I believe that at the school stages of learning mathematics, especially at the elementary level, it is quite possible to apply this advice.

My concern now is to point at some conflicts between *concept images* and *concept definitions* in the learning of mathematics in junior-high and high-school levels.

Let us start with some situations from the junior-high level. At this age, in many countries, the students study some chapters in Euclidean geometry. In this context, definitions are indispensible since very often new notions are introduced to the students, such as median, altitude, perpendicular bisector, and more. Also, some familiar notions, such as angles or parallel lines, for which the students have *concept images*, require certain clarification.

When learning new notions, there is no potential conflict between *concept definitions* and *concept images* that were formed before learning the definition. However, taking into account the fact that the students' minds (as well as our minds) tend to rely on *concept images* and not on *concept definitions*, we should do our best to form the correct *concept images* in the students' minds.

Take for example the concept of altitude of a triangle. Its definition is as follows: An altitude in a triangle is a line segment which is drawn from one vertex of the triangle and is perpendicular to the opposite side of this vertex or to its extension. However, a formal definition, generally, remains meaningless unless it is associated with some examples. The examples can be given by a teacher or by a textbook or they can be formed by the learners themselves. Unfortunately, quite often, in mathematical thinking, when a task is given to students, in order to carry it out, they consult their *concept image* and forget to consult the *concept definition*. It turns out that, in many cases, there are critical examples that shape the *concept image*. In some cases, these are the first examples introduced to the learner. In the case of the altitude, it is pedagogically reasonable to start with examples of altitudes in acute triangles. But, later on, in order to form the appropriate *concept image* of an altitude, the teacher, as well as the textbook, should give examples of altitudes from vertices of acute angles in obtuse triangles, and in right triangle. However, before this stage of the teaching takes place, the *concept image* of the altitude has been shaped by the stereotypical examples of altitudes in acute triangles (sometimes, even by the stereotypical examples of altitudes which are perpendicular to a horizontal side of a triangle). Thus, when the learners face a geometrical problem about altitudes which does not meet the stereotypes in their *concept image* they are stuck. It does not occur to them to consult the *concept definition* of the altitude, and if it does occur, they usually recall the first part of the definition ("a segment which is drawn from one vertex of the triangle, and is perpendicular to the opposite side of this vertex") and forget the additional phrase in the definition ("or to its extension"). Note that in a right triangle the altitudes from the vertices of acute angles are identical with the sides adjacent to the right angle.

Another example is the concept of median of a triangle. We should present it to the students in all kinds of triangle positions and not mainly in triangles in which the side to which the median is drawn is horizontal. If we use only the practice of drawing the median to the horizontal side of the triangle, we may find out that after a while, when the *concept definition* is forgotten or has become inactive, the student will have difficulty in identifying or drawing medians to sides that are not horizontal in triangles.

An additional example of this kind: At the junior-high level, in geometry, when a quadrangle is defined as a particular case of a polygon (a quadrangle is a polygon which has four sides), the learners have difficulties accepting that a concave quadrangle or a quadrangle that intersects itself are quadrangles too.

Anyhow, Euclidean geometry at the junior-high level is probably the best context to teach the role of definitions in a deductive structure. Here, students are expected to understand that the meaning of a concept is determined by its formal definition, and it does not matter what their previous views about the concept were. They are expected to play the game of mathematics as a deductive structure according to its rules. They are expected to follow the rules of the game. Some students may like this kind of game. Others may dislike it. Here, individual differences play a critical role.

It is similar to the fact that some people like basketball and others prefer football. We should respect individual differences and it is a pedagogical mistake to force changes in taste and inclinations. The differences have psychological reasons; some of them are structural, while some of them are acquired. As long as we are not concerned with moral issues, there is no justification for imposing on our students' games they do not like to play. We should be especially sensitive since mathematics, to a certain extent, is an obligatory discipline for all school students.

Sometimes, for the sake of "mathematical integrity," the curriculum includes topics for which the students do not have mathematical maturity or solid mathematical background: For instance, the case of irrational numbers. Some curriculums insist on introducing this concept to the students. They suggest the following definition: "An irrational number is one that is not rational," in other words, a number that cannot be expressed as a ratio between two integers (or as a simple fraction). A lot of mathematical ideas are required to understand this concept. The curriculum does not have the time to elaborate on it. The practice is to mention some irrational numbers, and the simplest practice at the junior high level is to mention π (the ratio between half the circle and its radius). At an earlier stage, in some countries, the students are told that π is approximately $\frac{22}{7}$. Thus, $\frac{22}{7}$ becomes a part of the *concept image* of irrational numbers in some students' minds. Thus, some students and some elementary teachers, when asked to give an example of an irrational number, point at $\frac{22}{7}$. The explanation is quite clear. The definition, which probably was too difficult to understand, was forgotten or ignored. When exposed to the notion of irrational numbers, the *concept images* became active, and thus, an example that so obviously contradicts the *concept definition* of an irrational number was given.

So far, I have dealt with the role of examples at the junior-high level.

With regard to algebra and calculus at the high-school level, my advice is to maintain an informal way of teaching. This was the way that mathematics was taught at the elementary level and algebra was developed on the ground of the mathematics that was presented to the students at the junior-high level. Changing this, all of a sudden, causes a discontinuity in the learning process. Generally speaking, discontinuities are not desirable since, as I claimed above, they require a conceptual change. Such a change may cause more students to become victims of mathematical difficulties.

A partial list of central concepts in algebra and calculus at the high-school level may include function, limit, derivative, continuity, and more. These concepts can be introduced by means of examples, which can be followed by general explanations. These explanations are a kind of informal definitions. Indeed, this approach may face some difficulties at certain intersections in future studies. In University courses for mathematics major students, if you do not introduce the Bourbaki definition of a function to the students, then they might not be able to deal with all kinds of weird functions presented to them in the curriculum. The Bourbaki definition of a function is as follows: A function f from a set X to a set Y (which is not necessarily different from X) is a subset of pairs (x, y), x is an element of X, y is an element of Y, such that if (x, y_1) is an element of this subset and (x, y_2) is also an element of it, then $y_1 = y_2$.

For junior-high and high-school levels, the recommended definition of a function is as follows: A function is a correspondence between a set X and a set Y (not necessarily different) which assigns to each element of X exactly one element of Y.

However, even if this definition is followed by some general examples (for instance, the correspondence which assigns to every animal its mother), eventually, the stereotypical *concept image* of a function is that of an algebraic formula. In Chapter 4, I described how this *concept image* was revealed by students' answers to the questionnaire I composed.

Why do some mathematicians prefer the Bourbaki definition to the last one? Because the last one uses the concept of a *correspondence*. The notion of *correspondence* relies on human actions, and according to some mathematicians, mathematical concepts should be independent of human actions. In order to avoid human actions, they introduced the Bourbaki definition.

There is also the question of whether to present all kinds of weird functions to high-school students. For instance, a function that assigns 1 to each rational number and assigns 0 to each irrational number. In my opinion, presenting these weird functions to high-school students does not add anything to the high-school students. These weird functions will be presented to some university students who have special love and talent for mathematics, should they decide to major in mathematics. This concludes my discussion on functions.

Now to the concept of the limit of a sequence. The formal definition of a limit of a sequence is as follows: L is the limit of the sequence a_n if for every positive small number ε there exists a natural number M such that for every n, if $n > M$ then $|a_n - L| < \varepsilon$.

No doubt that we do not want our high-school students to mess with such a monstrous definition. Therefore, this concept is usually acquired by means of examples. The generic example is the limit of the sequence $a_n = \frac{1}{n}$. The examples are usually followed by explanations like: "The elements of the sequence approach the limit (or tend or get closer to the limit) of the sequence." This explanation is a kind of informal definition. This is not an accurate definition, but it is usually sufficient for sequences which appear in high-school exercises.

However, the limit of the sequence $a_n = c$ needs further explanation because it does not fit the *concept image* formed by the generic example $a_n = \frac{1}{n}$, and by the informal definition of limit given above. Some students doubt whether this is a sequence because they are used to sequences in which the elements are different from each other or sequences in which the nth element depends on n. These students I tell that they can consider the nth element of the constant sequence as $a_n = (-1)^{2n}c$.

Another difficulty is accepting that the limit of the constant sequence is the constant itself. This difficulty lies in the idea (*concept image*) that the elements of a sequence are approaching its limit. They are getting close to it but they never reach it. The origin of this idea lies in the generic example mentioned above, the limit of $a_n = \frac{1}{n}$.

Here, again one can see the advantages and disadvantages of teaching concepts by means of examples. We can find ways to smooth the difficulty involved in the limit of the constant sequence by giving students additional explanations. We can say that in mathematics being close to a point does not necessarily mean that we are not at this point already. When we are at the point it is possible to say that we are close to it although this is not the common use of the expression "being close."

Similar advice can be given about the definition of a limit of a function. Here again the difficulty arises when we deal with the limit of the constant function. It turns out that the difficulties arise in the simplest case from the point of view of formal mathematics. Thus we see that formal mathematics simplicity is not necessarily a cognitive simplicity.

Last, but not least is the continuity of a function. In university mathematics, the continuity of a function is defined at a point. A function is continuous if it is continuous at every point that belongs to the domain in which it is defined. Again, it is recommended to avoid presenting a monstrous definition to the students. It is better to tell them that a function is continuous if it is possible to draw its graph without lifting the pen from the paper. This is not quite an accurate definition; there are continuous functions whose graphs cannot be drawn at all. They are presented to mathematics majors at the universities.

As I mentioned above, formal mathematics tries to avoid human actions like drawing or assigning elements in a correspondence. However, it is not the right thing for high-school mathematics. Very often when rigorous proofs are discussed in mathematics education forums, they are not recommended for the majority of high-school students. I would like to suggest that rigorous definitions are also not recommended for the decisive majority of high-school students.

The bottom line of this section is that in mathematical behavior we expect our students to consult the mathematical definitions (even if only an informal definition was given), and not only their *concept images*. They are expected to inhibit spontaneous reactions which are produced by their *concept images* before working on a given task. This is an additional reference to the title of my book—from intuition to inhibition.

Conjecture Formation and Conjecture Verification by Means of Examples in Common Everyday Thought Processes

Common everyday thought processes are developed in our minds before mathematical thought processes are imposed on them by formal school learning. Everyday thought processes are spontaneous, intuitive, and very often uncontrolled. However, they have enormous impact on our views about many important, as well as unimportant, matters of our world. When I say *world,* I mean the physical world as well as our society, culture, human beings, animals, etc. In everyday life, from early childhood, we are exposed to a lot of information about many topics. It is a fact that our minds develop views about these topics. Sometimes, these views are results of generalizations. We can assume that in our minds there exists a tendency to generalize. It can even be considered a need to generalize. In some cases, we are unaware of it. However, we can be driven by it without being aware of it.

Maslow (1970) considers "curiosity, cognitive impulses, the needs to know and to understand, the desires to organize, to analyze, to look for relations and meanings an essential part of human nature." If we accept this claim, we have enough justification to assume that the drive to generalize is also an essential part of human nature. Similar events lead us to form a generalization that forms a claim.

Let us call the generalization involved in the formation of claims a generalization schema. The generalization schema here acts in a similar way to the schema that is involved in concept formation; it involves a mechanism that identifies similarities and a mechanism that distinguishes differences.

Since I am trying to express myself with notions that are used in the philosophy of science (Popper, 1963), I will now use *conjecture* instead of *claim*. Thus, a generalization of events immediately attains the status of a "valid conjecture" in our minds.

It would be more correct to say "temporarily valid conjecture" because like all conjectures, at least theoretically, they can still be refuted. But usually the minute a generalization is made, it very often persists. Many people do not have the ability, as well as the willingness, to examine their generalizations and to change their minds about them.

It seems to me that it will be better if my above general explanations will be clarified by some examples. Let us consider very simple generalizations formed by little children:

Take, for instance, a little child who notices at a certain moment a barking of a dog. This child has already formed in his mind the concept of dog. After observing several different dogs that bark, he generalizes and concludes that dogs bark (or if you wish, the generalization: All dogs bark). Note that the concept of barking is also involved in this generalization.

As to generalizations formed by adults, I would like to illustrate them from my own experience. I have discovered that most generalizations occur to me when I come to new places as a tourist. Thought processes (like problem solving and generalizations) occur mainly upon arrival at new places and during the first few days of my stay there. I am supposed to go from one place to another, while I hardly know anything about transportation in the new place. I have to buy my food, and maybe additional things, while I have no information about supermarkets and other stores. In these situations, I try to rely on signs. Of course, I look for English signs. If I land at an airport in which English is not an official language, I have a problem. Yet, English signs also exist in countries where English is not an official language. So, the first thing that I have to figure out is whether there are English signs in the airport. I look around. Assume that there are no English signs. My goal is to get to a passport control point or to a suitable departure gate in case I am a transit passenger. Assume I discover that I am unable to rely on signs. Then, I try to ask somebody in English. Thus, I have to figure out whether people understand English. Assume I find out that people do not understand English. This is already a generalization. Another generalization was made earlier when I realize that I cannot rely on signs.

This generalization was made by a statistical inference from a small sample. However, in spite of that, it is possible that some people do speak English. How can I identify them? Perhaps, the people who do speak English are businessmen, who can be identified by their attire (wearing a tie) or by the bags they carry. Description of my thought processes can go on and on and therefore, before giving some additional examples, I would like to stop for a moment and make a comment. I am sure that everybody, from time to time, finds himself or herself in situations like the above in which they are supposed to use real-life problem-solving skills in order to survive. In these thought processes, very often generalizations are formed. I am also sure that everybody knows that some generalizations that we make turn out to be completely false. The important conclusion is: Be careful with generalizations. Sometimes, they are false.

Here are two additional examples of generalizations that I made as a tourist in Taipei, the capital of Taiwan. While walking on its streets, I did not notice any homeless people sitting or lying on the pavement, as one can see in New York City. So, I came to the conclusion that there are no homeless people in Taipei. Three days passed from the moment I reached this conclusion until I noticed one. Hence, I made the necessary changes to my first conclusion. Also, I did not notice any blonde Chinese women on the streets. All the blonde women that I noticed in Taipei were not Chinese. Thus, I generalized by claiming that there are no blonde Chinese women. This claim has not been refuted till now.

I give all these examples just to demonstrate how in everyday thought processes we are quite busy with drawing conclusions and particularly with making generalizations.

In some of the above examples, it is a question of survival (getting to a place in which you can spend the night and recover from an endless sequence of flights). In other cases, generalizations do not have any importance. However, does not the fact that we think about them support Maslow's claim that the needs to know, to understand, to organize, and to analyze are an essential part of the human nature?

Now, coming back to my main line of thought, I would like to say that, usually, after a generalization is made, it is quite natural for our minds to stop looking for counterexamples that may refute it. Moreover, very often we ignore counterexamples. It is crucial to emphasize that, in the majority of the cases, generalizations are intuitive. Therefore, they have similar characteristics to intuitions. Namely, they are immediate, spontaneous, rely on global impressions, and are not analytical (see Fischbein (1987)). This is the reason that generalizations quite often fail. In such cases, we use the notion overgeneralization in order to explain the failure.

In a children's book, I found the following story: The kindergarten teacher wants to comfort the child who cries desperately. "Eat this piece of chocolate and you will feel better." The child: "I don't like chocolate." The teacher: "In all my life I haven't met anybody who does not like chocolate. You must be kidding. Take it!" The child: "I don't like chocolate."

The above story is a typical example of how a person who has a certain idea about a certain topic refuses to accept a counterexample which contradicts his or her belief.

In an early work of mine (Vinner, 1997), I recommended that after forming a generalization, before rushing on with it, it is useful to stop for a while and examine it by analytical thought processes. Such a move in our thought processes can be considered a thinking tool. Such tools are extremely important for everybody. The question is in what context of the school curriculum should they be presented to the students, and how to make these tools a habit of the students' thinking. Is not mathematics an appropriate context for developing such thinking tools in the students?

Beliefs and Conjectures

Beliefs are conjectures that people refuse to give up. Everybody has some beliefs. Sometimes, it is useful to examine our beliefs since they might be wrong. It is important to understand how certain conjectures acquired the status of beliefs. We believe that the sun will rise tomorrow. We believe that if we drop something from our hands it will fall down. We believe that if we put our finger in a flame, the result will be a burn. I assume that all human beings share these beliefs. Therefore, I would like to call them universal beliefs. But there are other beliefs as well. There are beliefs which are not universal. What happens to us when we realize that one of our beliefs that we considered universal is not universal? Some of us think that people who do not share these specific beliefs with us are stupid or at least blind. However, some of us will be ready to examine these specific beliefs. Usually, we tend to admit that a specific belief of ours is wrong when we become aware of counterexamples. It turns out that admitting that a given example is a counterexample to one of our beliefs is not such a simple thought process. Sometimes, we make changes in our belief system such that the given example will not contradict it anymore. In other cases, we just refuse to be convinced that a given example is a counterexample. We reject it, explicitly or implicitly. For instance, some people believe in the evil eye, astrology, telepathy, aliens, and so on. (For additional examples, see Sagan (1995).) Other people believe that mathematics helps improve our general intellectual abilities (Poincare, 1914).

Those who do not believe in some of these and try to convince believers by counterexamples to abandon these beliefs know that the success rate of these attempts is very low.

As to us, mathematics educators, we stick to our belief that it is possible to teach mathematics in a meaningful way. This is in spite of the fact that there are infinitely many counterexamples.

And why don't we? Because this belief makes our acts as mathematics educators meaningful. We don't want to give up on our hope to teach mathematics in a meaningful way.

A Mathematical Example of a Wrong Conjecture

In some schools, computer micro-worlds are used to help students make conjectures and examine them. For instance, in the context of quadratic equations, one can lead students to the conjecture that the solutions of a quadratic equation of the form $ax^2 + bx + c = 0$, where a, b, and c are integers, should be divisors of c. There are infinitely many examples which support this conjecture. One of them is the equation: $x^2 - 5x + 6 = 0$. Its solutions are 2 and 3 and they are devisors of 6. Now, take the equation $x^2 + x - 5 = 0$. According to the above conjecture, the solutions of the equation should be among 5, −5, 1, and −1. However using the solution formula for the quadratic equation, we get:

$$x_1 = \frac{-1 + \sqrt{21}}{2}; \quad x_2 = \frac{-1 - \sqrt{21}}{2}.$$

This example can teach students to be cautious about their conjectures and to take into account that their conjectures may be wrong, not only in the mathematical domain.

The Importance of Learning Mathematics

I tried to add an ironical flavor to the title of this section. I am referring to Oscar Wilde's famous play *The Importance of being Ernest*.

Many people believe that mathematical knowledge is extremely important for every educated adult. In Chapter 1, I have already mentioned that I do not share this view. The fact is that the mathematical knowledge of the majority of people is very limited and they hardly use mathematics in their everyday life. Of course, there are some people who use mathematics in their professional life at a very high level and, generally speaking, mathematical knowledge is crucial to maintain science, technology, and medicine of our present generation. This does not imply that mathematics should be taught as a compulsory subject to the majority of the population. One reason that mathematics is taught to the extent it is taught is because it is a selection tool for further studies, as I mentioned in Chapter 2.

Nevertheless, since I do not think it is possible to change this practice, I recommend that we try to make the best out of it. Namely, while dealing with mathematical knowledge (i.e., while covering the syllabus), discuss also the thinking tools (mentioned already in this chapter in the section about conjecture formation and verification) concerning the need to examine our generalizations by analytical thought processes. It might be quite helpful for students to carry out some of these thought processes at crucial moments of their lives. These issues belong to the metacognition of mathematics learning. They require a reflective kind of teaching, which is, unfortunately, very rare.

Anyway, if there is any importance in learning mathematics, besides its being a selection tool for further studies, it is its potential to help students improve their thinking in some situations in everyday life.

Some readers may wonder if many sections of this chapter are relevant at all to mathematics education. Nevertheless, I consider them extremely relevant to mathematics education because I believe that there is no justification to teach mathematics to the decisive majority of students as an isolated subject. It should be related to everyday thought processes as well.

If mathematical thinking tools are not applied to additional domains, then the students' motivation to study mathematics will remain the same as it is today; namely, to score highly on the crucial examinations.

The Illusive Nature of Examples

In previous sections, I pointed out the potential danger of relying on examples as a solid ground for making generalizations and sticking to them. This danger also exists in everyday thought processes. I recommended there to examine generalizations and not to ignore counterexamples that may potentially refute them.

However, in everyday thought processes and in science, unlike mathematics, generalizations are very often the result of statistical inferences. With such generalizations, there is danger of over-reliance on counterexamples (single or very few). For instance, sometimes, when I discuss with colleagues and friends the importance of avoiding health risks such as smoking, being overweight, or lack of physical exercises, they point at counterexamples that had all these health risks and still reached the age of 90 years and more (Churchill, for instance). It turns out that when it is difficult for us to cope with some statistical generalizations, we somehow find ways to reject them. Here, the issue of rational thinking arises, and it will be discussed in Chapter 17.

Potentially Negative Influences of Mathematical Thinking on Thought Processes in Other Domains

There is a danger of adopting certain mathematical thought processes to thought processes in other domains. In the opinion of some mathematicians, it is unfortunate that not all intellectual domains behave like mathematics. Namely, not in all non-mathematical domains can concepts be defined by means of primary concepts and not all claims can be established by means of primary assumptions (axioms). The attempt to apply the model of mathematics to other domains sometimes has a negative impact. For instance, very often in everyday discussions, as well as in social-science discussions, people try to define the concepts they use and to impose their definitions on other people participating in the discussion. Quite often, they realize that their "definition" is not accepted by these other participants. This may cause, sometimes, unpleasant controversies and an unnecessary waste of time and intellectual energy. What is desirable in such situations is what we call in mathematics education "meaning negotiation." Meaning negotiation does not occur in mathematics. However, it is recommended in other contexts.

It is recommended not only because it is necessary for the sake of the discussion, but also because of the human relationships involved in intellectual discussions. Negotiations are required to solve conflicts. It is impossible to solve conflicts by unilateral stipulations. Moreover, in social sciences, as well as in everyday discussions, many statements do not have the status of a claim. They are suggestions of interpretation. The simplest and perhaps most common example that I can give here is the case of interpreting a slip of the tongue. In some cases of slips of the tongue, people suggest that they are Freudian slips of tongue and can be analyzed accordingly. However, there is no way to prove such a claim. It is a pseudoscientific claim, like many other claims in Freud's theory. On the other hand, it is not recommended, in my opinion, to dismiss such claims for not being scientific.

The Notion of Mathematical Proof

In addition to all that I have said above about the role of definitions and examples in forming mathematical concepts, we should remember that perhaps the most important mathematical concept, the concept of proof, does not have a definition (at least not at the high-school level and even not at the undergraduate level). In mathematical logic courses, there is a narrow definition of proof within what is called first-order predicate calculus. I do not want to describe it here because it is too technical and unfortunately, it does not cover the full notion of proof in mathematical theories. It is well reflected in Davis and Hersh (1981), pp. 39–40. It is a conversation between the university mathematics teacher, ironically called by the authors "the Ideal Mathematician" (I.M.) and a student (S.) who tries to find out what a mathematical proof is.

S.: Sir, what is a mathematical proof?
I.M.: You don't know that? What year are you in?
S.: Third-year graduate.
I.M.: Incredible! A proof is what you have been watching me do at the board three times a week for three years! That's what a proof is.
S.: Sorry, sir, I should have explained. I am in philosophy, not in math. I have never taken your course.
I. M.: Oh, well, in that case—you have taken some math, haven't you? You know the proof of the fundamental theorem of calculus—or the fundamental theorem of algebra?

S.: I have seen arguments in geometry and algebra and calculus that were called proofs. What I am asking you for isn't examples of proofs; it is a definition of proof. Otherwise, how can I tell what examples are correct?

I.M.: This whole thing was cleared by the logician Tarski, I guess, and some others, maybe Russell or Peano. Anyhow, what you do is, you write down the axioms of your theory in a formal language with a given list of symbols or alphabet. Then you write down the hypothesis of your theorem in the same symbolism. Then you show that you can transform the hypothesis step by step, using the rules of logic, till you get the conclusion. That's a proof.

S.: Really? That's amazing! I have taken elementary and advanced calculus, basic algebra, and topology, and I have never seen that done.

I.M.: Oh, of course, no one ever really does it. It would take forever! You just show that you could do it, that's sufficient.

S.: But even that does not sound like what was done in my courses and textbooks. So mathematicians don't do proofs after all.

I.M.: Of course we do! If a theorem isn't proved, it's nothing.

S.: Then what is a proof? If it's this thing with a formal language and transforming formulas, nobody ever proves anything. Do you have to know all about formal languages and formal logic before you can do a mathematical proof?

I.M: Of course not! The less you know, the better. That stuff is all abstract nonsense anyway.

S.: Then really what is a proof?

I.M.: Well, it's an argument that convinces someone who knows the subject.

S.: Someone who knows the subject? Then the definition of proof is subjective; it depends on particular persons. Before I can decide if something is a proof, I have to decide who the experts are. What does that have to do with proving things?

I.M.: No, no. There is nothing subjective about it! Everybody knows what a proof is. Just read some books, take courses from a competent mathematician, and you will catch on.

S.: Are you sure?

I.M.: Well—it is possible that you won't, if you don't have any aptitude for it. That can happen too.

S.: Then you decide what a proof is, and if I don't learn to decide in the same way, you decide I don't have any aptitude.

I.M.: If not me, then who?

It is quite clear from this imaginary entertaining conversation where the problem is. Students who later on become mathematicians, acquire the concept of proof by examples. The attempts of the mathematics teacher to define it fail one after another, thanks to the questions of the student. Thus, an important consequence can be concluded here: In many cases, there is no point in looking for definitions. Just rely on examples. The more examples you have encountered, the clearer the concept you are supposed to acquire will be. This is in fact the bottom line of the above conversation between the ideal mathematician and the student.

Chapter 10
Imagination Acts

In various learning situations, mathematics students are required to perform what I call "imagination acts." For instance: When studying Euclidean geometry they are required to think about a Euclidean point, which has no dimensions (no width and no length). This is in contrast to a physical point, which has dimensions, regardless of the sharpness of the pencil you use to draw it. Thus, when they perform a geometrical construction with a ruler and a compass they should bear in mind that what they draw is only a physical approximation of the "real" thing, the Euclidean configuration.

Moreover, we are asked to imagine a straight line which can be extended infinitely in both directions. However, in our physical world all straight lines are finite. When dealing with the density axiom (between any two points on a straight line there is another point and therefore, as a consequence, there are infinitely many points between any two given points on a straight line) we should bear in mind that the density of points on a Euclidean line is essentially different from the density of points on a physical line. In any given segment of a physical line, there are only a finite number of points.

Unfortunately, there are some students who are not capable of performing these imagination acts. These students should be advised to study a minimal amount of mathematics. We should be extremely careful not to consider them stupid or unintelligent. From my own experience, I can tell that these students can be very successful in history, literature, art, music, and many other domains in which mathematics is not required.

Some more examples of imagination acts which will be required in the next chapter: Can you imagine an infinite hotel whose rooms are across an infinite straight line (like the number line)? Can you imagine a group of infinitely many guests who come to stay in this hotel? If you cannot—just skip the chapter about the weirdness of infinity.

S. Vinner, *Mathematics, Education, and Other Endangered Species*, Mathematics in Mind, https://doi.org/10.1007/978-3-319-90035-3_10

Chapter 11
The Wonders of Infinity and Its Weirdness

In Which Set Do We Have More Elements?

When we start to deal with infinity, we very often pose the following questions to our students: "In which set are there more elements, in the set of all even numbers or in the set of all natural numbers? In which set are there more elements, in the set of prime numbers or in the set of all natural numbers?" Well, on the one hand the answers to the above questions are quite obvious. There are more elements in the set of all natural numbers because all the other above-mentioned sets are partial sets of the set of all natural numbers.

But let us attack the above problem from another angle. When we were first graders, very often we were asked to compare the sizes of two sets. For instance, a set of several pears and a set of several apples. Many students at this stage of their mathematical development can count. Hence, they count the pears, count the apples, and come up with the answer. However, we can ask them to reach an answer without counting. This requires an activity which was likely exercised at the beginning of their textbook, namely, the one-to-one correspondence. By a one-to-one correspondence between two sets we mean that with each element of the first set we associate one element of the second set, so that each element of the second set was associated to an element of the first set.

Definition: Two sets between which there is a one-to-one correspondence are called equivalent sets.

Adopting the above way to compare two sets, we can easily form a one-to-one correspondence between the above-mentioned partial sets of the natural numbers and the set of all natural numbers. If you observe the following two pairs of rows, you will easily be convinced that my claim is true.

S. Vinner, *Mathematics, Education, and Other Endangered Species*, Mathematics in Mind, https://doi.org/10.1007/978-3-319-90035-3_11

$$\begin{array}{cccccccccccc} 1, & 2, & 3, & 4, & 5, & 6, & 7, & 8, & 9, & 10, & \dots \\ 2, & 4, & 6, & 8, & 10, & 12, & 14, & 16, & 18, & 20, & \dots \end{array}$$

$$\begin{array}{ccccccccccccc} 1, & 2, & 3, & 4, & 5, & 6, & 7, & 8, & 9, & 10, & 11, & 12, & \dots \\ 1, & 2, & 5, & 7, & 11, & 13, & 17, & 19, & 23, & 29, & 31, & 37, & \dots \end{array}$$

Moreover, by the same method one can show that there are as many powers of millions as natural numbers:

$$\begin{array}{ccccccccccc} 1, & 2, & 3, & 4, & 5, & 6, & 7, & 8, & 9, & 10, & \dots \\ 10^6, & 10^{12}, & 10^{18}, & 10^{24}, & 10^{30}, & 10^{36}, & 10^{42}, & 10^{48}, & 10^{54}, & 10^{60}, & \dots \end{array}$$

Weird, isn't it? However, people get used to weird things. It might take some time. Just recall when you were told that your belief about the sun making half a circle in the sky from east to west during the day was wrong. Probably, you have forgotten it, having gotten used to the true explanation, which was likely presented to you in the third grade.

Now, to an even more confusing question about infinite sets: Are the set of rational numbers (the set of all simple fractions, including improper fractions, in case you are not familiar with this notion), and the set of natural numbers equivalent? In other words, is there a one-to-one correspondence between the set of all rational numbers and the set of natural numbers? At first sight, the impression is that such a correspondence does not exist.

Nevertheless, we are going to present such a correspondence to the reader.

Think of all the reduced fractions whose denominator and numerator have a sum of 1. There is only one such fraction: $\frac{0}{1}$. In the following steps, we will not consider 0 anymore. Now consider all the reduced fractions whose denominator and numerator have a sum of 2. In this case, there is also only one such fraction: $\frac{1}{1}$. So, until now we have counted 2 fractions. We write them in a row. In the next step, we consider all the reduced fractions whose denominator and numerator have a sum of 3. Here, we already have two fractions: $\frac{1}{2}$, $\frac{2}{1}$. We add them to the row. In the next step, we consider all the reduced fractions whose denominator and numerator have a sum of 4. They are: $\frac{1}{3}$, $\frac{3}{1}$. And what about 5? Here we get: $\frac{1}{4}$, $\frac{2}{3}$, $\frac{3}{2}$, $\frac{4}{1}$. (Note that there is a tendency not to consider natural numbers fractions. However, natural numbers are fractions whose denominator is 1.)

In each step, we keep adding the fractions at the end of the row. When you examine the case of 6, for instance, you get: $\frac{1}{5}$, $\frac{2}{4}$, $\frac{3}{3}$, $\frac{4}{2}$, $\frac{5}{1}$. In the above sequence, we got three non-reduced fractions: $\frac{2}{4}$, $\frac{3}{3}$, $\frac{4}{2}$. They are equal to $\frac{1}{2}$, 1, and 2, respectively. They were counted already in a previous step. Therefore, we will not count them again.

At this stage, I hope that the reader understands how to proceed. Thus, we will describe the nth step. In this step, we consider all the reduced fractions whose denominator and numerator have a sum of n. They are among the following fractions: $\frac{1}{n-1}$, $\frac{2}{n-2}$, ..., $\frac{(n-1)}{1}$. As we showed above, in this sequence we can find fractions that are not reduced and which have been counted in a previous step. We shall omit them. Thus, in the nth step we have at most $n - 1$ reduced fraction which we add to our row. For instance, the row which is formed by steps 1, 2, 3, 4, 5, 6 above is: 0, 1, $\frac{1}{2}$, 2, $\frac{1}{3}$, 3, $\frac{1}{4}$, $\frac{2}{3}$, $\frac{3}{2}$, 4, $\frac{1}{5}$, 5.

Although there is an infinite number of steps in this procedure, at the end of every step we are left with a finite number of fractions which are added to the row. If we write this row under the row of natural numbers, we can see the one-to-one correspondence between the simple fractions, or rational numbers, and the natural numbers.

More Weird Situations

Let us imagine an infinite hotel, whose rooms are arranged along an infinite straight half line. This hotel was the idea of the great mathematician David Hilbert. Assume also that the sign on the front gate says: "No vacancies." Around midnight a guest comes and asks for a room. "Sir, I am terribly sorry there is not a single room left," says the receptionist. "Don't worry. I can arrange it. I am a mathematician."

"How?" asks the receptionist.

"It is very simple," says the mathematician. "Tell the guest in the first room to get out of his room, to knock on the door of the next room, and to ask the guest to move to the next room, and so on. Thus, the first room will become vacant, and I can stay in it."

"Amazing!" says the receptionist.

Imagine now a slightly different situation: A day later, again, around midnight a finite group of people come to the same hotel on the front gate of which there is a sign "No Vacancies." "We would like to stay in your hotel," they say. "I am terribly sorry," says the receptionist. "If you were only one, I would know how to arrange it. A brilliant mathematician taught me. He is still staying in the first room. However, since you are many it cannot be arranged." "Don't worry," says somebody in the group. "I am a mathematician and I will teach you how to arrange it."

"How?" wonders the receptionist.

"OK," says the mathematician, "first of all, ask your guests to perform the same procedure they performed yesterday. Thus, you will have vacancy in the first room. At the next step, ask them to perform the same procedure again. After performing this procedure you will have vacancies in the first two rooms. In a similar way, ask them to repeat this procedure n times, where n is the number of people in our group."

"Amazing!!" says the receptionist. He is very pleased to receive the accommodation fees from n additional guests and goes to bed quite satisfied.

A day later, around midnight, an infinite group of people arrives to the same hotel and asks the receptionist to stay in his hotel. The sign at the front door says: "no vacancies." At this stage, the receptionist becomes hysterical. "Look," he says to the group, "if you were a finite group I'd know how to arrange it, but since your group is infinite, it simply cannot be arranged." "Don't worry," says one of the group, "I am a mathematician, and I will tell you how to arrange it."

"I can't believe you," says the receptionist with tears in his eyes.

"Listen to me," says the mathematician. "Just tell every guest in your hotel to move into the room whose number is two times the number of their present room. Then all the odd-numbered rooms will be available, and you will have infinitely many vacant rooms as required."

"Amazing!" Says the receptionist. "You, mathematicians, you are magicians! People should really adore you and your wonderful invention, the infinity. This invention formed this infinite hotel and made my boss infinitely rich. It also made me quite rich."

Note that, in order to follow the above story, you were required to use what, in Chapter 10, I called "imagination acts." I do not believe that this was too hard for you, unless you do not want to mess with such stories, which might be considered ridiculous.

And What About the Real Numbers?

Let us move now to more serious business.

By now we have proved that some infinite sets (including the set of all fractions) are equivalent to the set of all natural numbers.

In order to answer the same question about the real numbers, we have to clarify some additional facts about rational numbers.

First, we will deal with the conversion of fractions to decimals. By the way, sixth grader Benny was involved in this activity in the classic paper of Erlwanger (1973).

When this topic is first presented to students, we start with simple fractions like $\frac{1}{2}$, $\frac{1}{4}$, $\frac{1}{5}$, etc. This is usually a sixth-grade activity and it is presented to the students after they learn long division. Using long division in order to convert a fraction into a decimal can give two possible results as described in the following sections.

Finite Decimals

For instance, the above-mentioned fractions: $\frac{1}{2} = 0.5$, $\frac{1}{4} = 0.25$, $\frac{1}{5} = 0.2$. More examples: $\frac{1}{8} = 0.125$, $\frac{1}{400} = 0.0025$.

A generalization of the above examples can be given by using the language of algebra. We shall state it by the following theorem:

A given fraction $\frac{m}{n}$ (where m and n are natural numbers) is equal to a *finite decimal* if and only if the denominator n has the form $2^i \cdot 5^j$ where i and j are natural numbers. (One of them or even both of them can be zero. Just to remind you: $a^0 = 1$. If both i and j equal zero, then the denominator n is equal to 1 and we deal with a natural number m.)

Infinite Decimals

The simplest example of an infinite decimal is $\frac{1}{3}$. Using long division, we get $0.333\ldots$ where the digit 3 recurs indefinitely. I will denote a recurring decimal by putting a horizontal line above the recurring digits, like: $0.333\ldots = 0.\bar{3}$.

The readers who want to carry out the calculations are advised not to use their calculators. Calculators are good for all practical purpose but they are not accurate from the mathematics point of view. Calculators give you only a finite number of digits. For instance, my graphic calculator tells me that $\frac{1}{3} = 0.3333333333$.

Thus, we have finished dealing with the simplest case that long division does not end. Let us move now to another case. It will be represented by $\frac{1}{7}$. If you carry out 10 steps of the long division, you will get 0.1428571428. Then, you notice that after 6 steps the digits start to repeat in the previous order. Actually, we will get an infinite decimal with infinite repetitions of the 6 following digits: 142857. That is, $0.\overline{142857}$ according to the above notation. It is not an accidental result. It happens because of the following reason: When you divide by 7 using the long division algorithm, you can get only 1, 2, 3, 4, 5, and 6 as remainders (the remainder is always smaller than the denominator.) Therefore, after at most 6 steps one of the remainders must recur. The moment it recurs the previous digits will reappear in the same order. (We said "at most" because it can happen earlier. For instance, with $\frac{1}{9}$ it happens after one step $\frac{1}{9} = 0.\bar{1}$. With $\frac{1}{11}$ it happens after 2 steps $\left(\frac{1}{11} = 0.\overline{09}\right)$ and with $\frac{1}{13}$ it happens after 6 steps $\left(\frac{1}{13} = 0.\overline{076923}\right)$.) In order to follow my above claims, you can use your calculators. However, if you have a cruel teacher who asks you to find the infinite recurring decimal for $\frac{1}{23}$ then using your calculator will be useless because the digits start repeating after 22 steps.

The type of the above infinite decimals is called a recurring decimal.

We can sum up what we have seen in the last two sections: Every simple fraction can be represented by a decimal—a finite decimal if the denominator is of the form $2^i \cdot 5^j$, and an infinite recurring decimal if the denominator is not of the form $2^i \cdot 5^j$.

Let us deal now with the opposite problem. Assume an infinite recurring decimal is presented to you. How will you find the simple fraction which is equal to it? I will explain it with some examples.

Let us start with a very simple example like: $0.111\ldots$, namely, $0.\bar{1}$.

Denote the above recurring decimal by y. Namely, $y = 0.\bar{1}$.

Now, multiply both sides of the above equality by 10. We get:

$$10y = 10 \times 0.\bar{1} = 10 \times 0.111\ldots = 1.111\ldots = 1 + 0.111\ldots = 1 + 0.\bar{1} = 1 + y.$$

Thus, we have obtained: $10y = 1 + y$ or $9y = 1$; $y = \frac{1}{9}$, namely, $0.\bar{1} = \frac{1}{9}$. (We already knew that from the previous discussion.)

Using the same procedure, we can continue playing with all similar recurring decimals $0.\bar{2}$, $0.\bar{3}$, $0.\bar{4}$, $0.\bar{5}$, $0.\bar{6}$, $0.\bar{7}$, $0.\bar{8}$, $0.\bar{9}$. We get: $0.\bar{2} = \frac{2}{9}$, $0.\bar{3} = \frac{1}{3}$

(we have seen it before), $0.\bar{4} = \frac{4}{9}$, $0.\bar{5} = \frac{5}{9}$, $0.\bar{6} = \frac{6}{9} = \frac{2}{3}$, $0.\bar{7} = \frac{7}{9}$, $0.\bar{8} = \frac{8}{9}$. But what about $0.\bar{9}$?

We can now use the same method for $0.\bar{9}$: Denote $0.\bar{9}$ by y. Namely, $y = 0.\bar{9}$. Multiply this equality by 10. We get:

$$10y = 10 \times 0.\bar{9} = 10 \times 0.999\ldots = 9.999\ldots = 9 + 0.999\ldots = 9 + 0.\bar{9} = 9 + y.$$

Therefore: $9y = 9$, namely, $y = \frac{9}{9} = 1$. This shows that $0.\bar{9} = \frac{9}{9} = 1$.

Surprisingly enough the recurring decimal $0.\bar{9}$ is equal to 1.

However, the recurring decimal $0.\bar{9}$ is a legitimate creature in itself.

We will bear this in mind for further discussion but in the meantime let us deal with a more complicated case.

Take for instance the recurring decimal $0.\overline{435}$. Denote it by y. Namely: $y = 0.\overline{435}$. Multiply both sides of the above equality by 1000. We get:

$$\begin{aligned} 1000y &= 1000 \times 0.\overline{435} = 1000 \times 0.435435\ldots \\ &= 435.435\ldots = 435 + 0.\overline{435} = 435 + y. \end{aligned}$$

Therefore: $999y = 435$.

$$y = \frac{435}{999} = \frac{145}{333}.$$

The readers are advised to check the above calculations with their calculators.

Of course, there are more complicated cases in which the recurring part of the infinite decimal does not start immediately after the decimal point. For instance: $0.73435435435\ldots = 0.73\overline{435}$.

Readers who like intellectual challenges are invited to cope with this case. By now, they have all the mathematical tools to cope with it. However, I do not want to involve the majority of my readers in the above complicated case.

At this stage, I would like to present to you three reflective comments:

1. Note that in order to follow the above procedures one needs to have a great amount of concentration ability. He or she should focus on the topic for a relatively long time. They should keep in their minds the line of thought presented in the procedures. Not everybody can do it and perhaps this ability is one of the distinctions between people who have a mathematical mind and those who do not have it. If you do not have it, do not feel bad about it. Probably, you can be good in other domains. Just study the minimal amount of mathematics which is required to be accepted for further learning (matriculation, SAT, etc.). There are many academic domains in which mathematics is not involved (literature, history, art, law, psychology, medicine, and more). My English literature teacher in high school told me after I graduated that when he was a high-school student he was a failure in mathematics. However, I will never forget the lessons in which I studied with him the Shakespearean plays *Julius Caesar*, *Macbeth,* and *Hamlet*.

2. Note that in order to establish my claim that every recurring decimal is equal to a simple fraction I avoided a general discussion. I have established my claim by means of examples. I have done this because of two reasons: (a) To establish it in a general way would be extremely complicated. It would require a massive use of the language of algebra and by doing it I might lose my readers. (b) The examples are quite convincing. Therefore, the preferable pedagogy is to avoid complicated explanations even if rigorous mathematicians like them. Our goal as mathematics teachers is to ease our students' lives and not necessarily to please rigorous mathematicians.
3. There are cases during the course of mathematics teaching in which we do not have general means to present some notions. One of these notions is the notion of proof. It was discussed in the last section of Chapter 9.

Two Decimal Representations of Simple Fractions with Denominator of the Form $2^i \cdot 5^j$

From the above discussions, we have learned that if the denominator of a simple fraction has the form $2^i \cdot 5^j$ it can be represented as a finite decimal, for instance: $\frac{1}{4} = 0.25$.

It turns out that these simple fractions can also be represented by an infinite recurring decimal, for instance: $\frac{1}{4} = 0.24\bar{9}$. The following calculation shows how from $\frac{1}{4}$ we get $0.24\bar{9}$:

$$\frac{1}{4} = 0.25 = \frac{25}{100} = \frac{24}{100} + \frac{1}{100} = \frac{24}{100} + \frac{0.\bar{9}}{100} = 0.24 + 0.00\bar{9} = 0.24\bar{9}.$$

Note that in the above calculation we used the equality $1 = 0.\bar{9}$, which we saw in the previous section, to replace the nominator 1 by $0.\bar{9}$.

In fact, every simple fraction with denominator of the form $2^i \cdot 5^j$ has two representations—a finite decimal and an infinite recurring decimal.

In the following discussion, we would like to avoid the situation of two representations for a number, and therefore we decide to represent all simple fractions only by their infinite recurring decimals.

Note that every rational number has a single representation by a recurring infinite decimal. Specifically, if two infinite decimals have different digits in a certain place, then these two infinite decimals are different numbers. For instance, consider the two infinite decimals 0.2389754056... and 0.2388754056.... They are different at the fourth place after the decimal point. The difference at the fourth place makes them different. Just subtract the second from the first and you will get 0.0001.

The bottom line that we should remember for the following discussion: Every rational number can be uniquely represented as an infinite recurring decimal.

The Real Numbers

The notion of real numbers is mentioned sometimes during the course of learning in high school. However, it does not get a meaningful treatment. This is because common tests do not deal with concepts. Mainly, they deal with procedures. Examiners wish to see whether the students can use certain procedures which they studied in order to solve routine questions.

However, the real numbers include all simple fractions, that is, the rational numbers, as well as other numbers which are called irrational numbers.

At the end of the previous section, we agreed to represent every rational number by an infinite recurring decimal. Now, the irrational numbers will be represented by infinite non-recurring decimals. This implies that every real number is an infinite decimal (recurring or non-recurring).

Among the irrational numbers there are two famous numbers: π and the square root of 2. Usually, when mathematics teachers mention that π is irrational they do not prove it (it is proved only in Number Theory in graduate mathematics courses.) As to the square root of two, in some high-school mathematics classes the teachers bother to prove that there is no fraction the square of which is equal to 2. In other words, the square root of 2 is irrational.

By the way, the notion irrational means that this number is not a ratio of two integers. It has nothing to do with the concept of rational in the sense of logical or reasonable.

As already mentioned, the fraction $\frac{22}{7}$ is a good approximation of π. $\frac{22}{7} = 3.142857142857\ldots = 3.\overline{142857}$ is an infinite recurring decimal, whereas $\pi = 3.141592654\ldots$ is an infinite decimal, but without recurring digits (because otherwise it would mean that π is rational). (As mentioned in the section about *concept images* in mathematics in Chapter 9, the fraction $\frac{22}{7}$ becomes part of the *concept image* of π, and therefore some students give the answer $\frac{22}{7}$ when asked to give an example of an irrational number.)

Although during the course of high-school learning only two irrational numbers are mentioned, there are infinitely many irrational numbers. All the square roots of the prime numbers are irrational (the proof of this claim is quite similar to the proof that the square root of 2 is irrational. This proof will be given in Chapter 14.) In fact, all square roots of non-square numbers, such as 10, 11, and 12 are irrational.

Since we proved that the sets of even numbers, prime numbers, some other sets that we mentioned earlier, and especially the set of all rational numbers are all equivalent to the set of natural numbers, it may seem to us that the set of real numbers is also equivalent to the set of all natural numbers. It may even seem to us that all the infinite sets of numbers are equivalent. However, this intuition is wrong, as we will see in the next section.

The Set of Real Numbers Is Not Equivalent to the Set of Natural Numbers

First, I am going to prove the claim: The set of all real numbers between 0 and 1 is not equivalent to the set of all natural numbers. Recall that every real number is an infinite decimal (recurring or non-recurring). The real numbers between 0 and 1 have 0 to the left of the decimal point.

We will prove it by contradiction. (This will be the first example of a proof by contradiction, a notion which will be discussed in Chapter 14.)

Our proof will start with the following suggestion: Assume that the set of all real numbers between 0 and 1 is equivalent to the set of all natural numbers. (Such an assumption is, as a matter of fact, the beginning of all proofs by contradiction. We ask ourselves to assume that our claim is not true.) Hence, if we assume that the set of all the real numbers between 0 and 1 is equivalent to the set of all natural numbers, then there is a one-to-one correspondence between these sets. This means that we can write down a pair of columns, similar to the pairs of rows we had in the first section of this chapter, so that in the left column we have all the natural numbers, and in the right column we have all the real numbers between 0 and 1 one below the other. Thus, on the right, we have an infinite column of rows, and each row is of infinite length (the digits of the infinite decimal in that row):

$$\begin{array}{ll} 1 & 0.143345836659\ldots \\ 2 & 0.577733275777\ldots \\ 3 & 0.476473565398\ldots \\ 4 & 0.333333333333\ldots \\ 5 & 0.897897897897\ldots \\ 6 & 0.999999999999\ldots \\ \vdots & \vdots \end{array}$$

At this stage, we have overcome the first obstacle. Now, let us attack the next obstacle. The next obstacle is a notational one. We will focus on the right column of real numbers, on the digits to the right of the decimal point. We would like to denote a digit that appears in a certain place of a certain row in our column. In order to do it, we will need two indices. For instance, to denote a digit which appears in the third row in the fifth place we will have to use indices 3 and 5. The digit that stands there will be denoted $a_{3,5}$. In general, the digit which stands in the ith row in the jth place will be denoted by $a_{i,j}$. Now we can express in symbols what we described above when we said that we can write down all the elements of the set of real numbers between 0 and 1. Here is the result (remember, all the a's with the indices in each row, are digits of the real number in that row):

$$
\begin{array}{ll}
0.a_{1,1}a_{1,2}a_{1,3}\ldots\ldots a_{1,n}\ldots\ldots & \text{(this is the first row)}\\
0.a_{2,1}a_{2,2}a_{2,3}\ldots\ldots a_{2,n}\ldots\ldots & \text{(this is the second row)}\\
0.a_{3,1}a_{3,2}a_{3,3}\ldots\ldots a_{3,n}\ldots\ldots & \text{(this is the third row)}\\
\vdots & \\
0.a_{j,1}a_{j,2}a_{j,3}\ldots\ldots a_{j,n}\ldots\ldots & \text{(this is the } j\text{th row)}\\
\vdots &
\end{array}
$$

At this point, we have got an infinite square. Again, in order to understand it one has to rely on imagination act (recall Chapter 10).

Let us write down a zero and a decimal point and continue writing all the digits of the diagonal of this infinite square (starting from its upper left corner). We get: $0.a_{1,1}a_{2,2}a_{3,3}\ldots\ldots a_{n,n}\ldots\ldots$, which is an infinite decimal.

Now, change each digit of this decimal to a different digit. By doing this we get an infinite decimal which is different from all the infinite decimals in the above infinite square (and therefore is not included in the column).

At this point, we have reached a contradiction. Why a contradiction? Because we assumed that we could write down in a column all the infinite decimals between 0 and 1 and just now we have pointed at an infinite decimal that is not included in the column. "Only one?", you may ask. "Then why not add it to the list, let us say, at the beginning?" The answer is that if you do that we can repeat the entire procedure again and obtain another infinite decimal which is not on the list.

Q.E.D.

From the above proof it turns out that in the set of real numbers between 0 and 1 there are too many elements to be all included in a one-to-one correspondence with the natural numbers. Therefore, the same is certainly true for the set of all real numbers, which has even more elements than the set of real numbers between 0 and 1. This establishes the claim that the set of real numbers and the set of natural numbers are not equivalent.

Well, I assume that the above proof was not easy to follow. It required, as I mentioned above, to be focused on a long sequence of arguments. Also, you have an additional difficulty, which is the notational one. Because of that, mathematics is not a simple domain. It requires good will and persistence. Perhaps these are also properties which characterize the mathematical mind.

The bottom line of this section is that not all infinite sets are equivalent. This result was the major mathematical achievement of one of the greatest mathematicians of all generations, Georg Cantor (1845–1918). He was the one who developed set theory. His theory was not accepted immediately by some great mathematicians and, as a result, Cantor suffered a lot. He suffered chronic depression and spent some years in psychiatric hospitals. At the end of his life, he received the recognition he deserved but he died in poverty and malnourishment. Thus, if you have difficulties accepting what I presented to you above—just know that you are in a company of some great mathematicians.

Chapter 12
Mathematics, Educational Values, and Intellectual Values

The emphasis on values is an essential point in my approach to teaching mathematics. Here, I call for the elimination of the distinction between teaching and educating. Teaching is conceived as covering the syllabus and preparing the students for certain exams. Educating means recommending to students certain modes of thinking, and certain modes of behavior that are appropriate for the human society from the moral point of view.

Moral Behavior: The Golden Rule

In my lifelong career as a mathematics educator, I have met many teachers who refuse to undertake the role of the educator. Some of them even tried to justify it by saying: "Who are we to tell somebody how to behave?"

My answer to them was: "If you want to avoid telling your students how to behave you are in the wrong profession." It is quite clear that parents are supposed to tell their children how to behave. It should also be the teacher's task, beyond telling their students to behave properly in class. In their classes, the teacher tells the students, for instance, not to disturb the teacher, to pay attention to him or her, to listen to their friends when they answer the teacher's question, and so on. However, this is only a small part of education. The main part is to recommend to them how to behave in all possible situations in their life, at school, and outside school. This can be done by what people call "the golden rule." There are many versions of this rule in different cultures.

My preferred version is that of the first century Jewish sage Hillel, who advised: "What you hate—do not do to your friend." From here the way to discuss with students even philosophical issues as the meaning of life is quite short.

Real educators, so I believe, should also point out to their students certain directions in their adult life. For instance, at the upper level of high school some students start to think what profession they will choose after finishing. Will they be

S. Vinner, *Mathematics, Education, and Other Endangered Species*,
Mathematics in Mind, https://doi.org/10.1007/978-3-319-90035-3_12

medical doctors? Will they be teachers, nurses, lawyers, politicians, etc.? And whatever they choose to be, how will their choice serve their society or their selfish goals like making money or becoming famous or influential? Of course, I do not ignore the natural wish of many people to become rich, famous, and influential ("Come to us and we will make you rich and famous," was an advertisement which I saw several years ago of a certain university in order to attract students to study there.) However, as an educator I recommend to my students to choose the altruistic direction for their life.

The Value of Procedures

One of the values that can be learned from mathematics is procedures. There are a lot of procedures in mathematical activities. Generally speaking, and not necessarily in mathematics, a procedure is a sequence of instructions that tells you how to behave. Here are some mathematical examples:

The Order of Arithmetic Operations Makes a Difference

In the exercise $13 + 9 \times 2 - 10 + 11 \times 3$, we all know that we should calculate first the multiplications and only later on the additions and subtractions. If you do that the result will be 54. If you ignore the instruction and you calculate the operations in the order in which they appear you will get: $((((13 + 9) \times 2) - 10) + 11) \times 3 = 135$. A mistake!

In Some Procedures the Order of Simplification Does Not Make a Difference

For instance: When we have to solve a quadratic equation like: $\frac{2}{x+1} - \frac{5}{x+2} = 1$ we can multiply both sides by the common denominator of the left-hand side and get:

$$\text{(I)} \quad 2(x + 2) - 5(x + 1) = (x + 1)(x + 2).$$

Or we can focus on the left-hand side of the given equation, rewrite it with a common denominator and get:

$$\text{(II)} \quad \frac{2(x+2) - 5(x+1)}{(x+1)(x+2)} = 1.$$

In both ways, we will get $x^2 + 6x + 3 = 0$ as the canonical form of the given equation.

Doing (I) seems simpler to me but still you can proceed the way you choose.

Procedures in Everyday Behavior

So, the educational message of the mathematics teacher is to respect procedures while doing mathematics because not respecting them will end up in a low grade. However, the recommendation should be to respect procedures outside mathematics as well.

When you get outside mathematics, there are hundreds of procedures. Some of them are dictated by law. For instance, it is forbidden for a driver to cross an intersection through a red light. If he does, he may be lucky, and get away with it. If he is less lucky, he may be caught by a policeman. And if he is even less lucky, he may cause a terrible accident, get killed, or kill other people. Yet, there are drivers who cross intersections through a red light. They ignore the driving lessons they had before getting their drivers' licenses.

There are instructions for pedestrians as well as for drivers concerning crosswalks. The instructions for drivers are to slow down when approaching a crosswalk and to stop if they see a pedestrian intending to cross the street.

The instructions for pedestrians are to look both ways and to wait until all approaching cars stop and all lanes of the crosswalk are clear.

However, I read on a sign on the New York City Subway that 70% of the pedestrians who were killed in New York City that year, were killed on a crosswalk. Thus, again, not following procedures promptly can cost us our lives, or lives of other people.

Other procedures that can avoid casualties are listening to warnings, like bad weather warnings, and following related safety rules.

More procedures: When entering our cars, by law, we have to fasten our seat belts. When should we do that? Before turning on the engine or after? It does not matter. Before starting our drive or after? It does matter—doing it while driving is dangerous because we cannot keep both our hands on the steering wheel.

More procedures: When leaving our office or our home, it is recommended to turn off all the lights and the air conditioning. In some offices, you can find this request at the exit. Not doing this will increase our expenses, but it is not dangerous.

When entering a concert hall or a theater, we are asked to turn off our phones. In some places, it is even announced on loudspeakers. However, how many times have we heard phones ringing in the middle of a show?

Now, as to individual procedures: There are hundreds of them. For instance, when you take a shower, do you prefer to brush your teeth before or after? When you get dressed, is it better to put on your shirt first or your trousers first? Note that in some procedures the order makes a difference. You cannot put on your shoes before putting on your socks.

As to me, during the day I find myself quite busy deciding which procedure is better in order to obtain a certain result. Sometimes I get tired of this. Then, I say to myself: "Give me a break!"

The Value of Analytical Thinking Outside Mathematics

This value was discussed in the section titled "The Importance of Learning Mathematics" in Chapter 9. It is one of the most important values that can be taught in the mathematics classroom. Specifically, pointing out to students that analytical thinking used in mathematics, should also be used in non-mathematical contexts.

Chapter 13
What Are Elementary Mathematics Teachers Expected to Know?

Elementary mathematics teachers should know which mathematics topics are required in everyday life and should of course know these topics. These topics are: Counting, adding and subtracting, multiplying, dividing, calculating change, calculating tips, and percentages. They should also know that mathematics is used as a filter for further studies.

Many mathematics teachers, especially at the secondary level, consider students who have difficulties in mathematics stupid or unintelligent. It is an arrogant approach. At that stage in high school, some prospective elementary mathematics teachers develop negative attitudes toward mathematics. However, if we accept what I have just said about the role of mathematics in everyday life and about its being a filter for further studies then we should adopt a more humble view about the status of mathematics. But before continuing our discussion on what elementary mathematics teachers are expected to know let us deal with the question of who the elementary mathematics teachers are.

Typical Profiles of Elementary Mathematics Teachers

In order to answer this question, we have to have some information about the teachers at the elementary level. In order to get such information, we should conduct an international survey. However, conducting such a survey is almost a mission impossible. Hence, what I say here relies on impressions I got in my country and in other countries where I have met elementary mathematics teachers and spoken to teacher trainers.

Hence, relying on these impressions, what are the typical profiles of elementary teachers? Well, elementary teachers have many profiles. I have chosen to present some anecdotal impressions of elementary teachers in my country and in the USA.

I have asked elementary school teachers (especially in grades 1–3) in my country and in the USA why they decided to become elementary school teachers.

S. Vinner, *Mathematics, Education, and Other Endangered Species*, Mathematics in Mind, https://doi.org/10.1007/978-3-319-90035-3_13

Their answer was usually that they liked the human interaction with little children. Being involved with the children's intellectual and emotional development gives them a lot of satisfaction.

Usually, these teachers do not score very high on the college entrance examinations. For me, mentioning it is quite problematic. It is problematic because I cannot ignore the fact that this claim is expressed by some people with arrogance. It is also problematic because it suggests forming hierarchies of people by comparing their scores on college entrance exams. I myself do not believe in measuring people. The title of a canonical book on this issue is *The Mismeasure of Man* (Gould, 1981). To the educational community, I would like to suggest even a stronger title: *The Immeasurable Man*. Modern psychologists speak about many kinds of intelligence: emotional intelligence, social intelligence, and more (Gardner, 1993; Goleman, 1995). These kinds of intelligence are not less important to the success of teachers than their cognitive intelligence. Yet, there do not exist tests to measure these kinds of intelligence. In spite of all the above, I understand the need to evaluate someone's ability to study a certain domain when he or she wants to study it at a college. Since mathematics is an important component of college entrance exams, the fact that prospective elementary teachers do not score highly on college entrance exams may predict that they will have some difficulties studying certain mathematical topics. Observing pre-service and in-service courses reveal quite often the mathematical weaknesses of some pre-service and in-service elementary teachers. For example, I am told by colleagues who teach remedial courses to these teachers that a significant number of them have difficulties in solving word problems like the following:

1. David holds $\frac{5}{8}$ of the shares of a certain factory. He gives his son Daniel $\frac{2}{3}$ of his shares. What part of the factory shares is owned by Daniel after this transaction?
2. Barbecuing meat causes it to lose $\frac{1}{5}$ of its weight. What was the original weight of a piece of meat if after barbecuing its weight was 300 g?

The Knowledge Domains Which Elementary Mathematics Teachers Are Supposed to Have

I believe that now, all over the world, teacher training has adopted the approach that the domains of content knowledge and of pedagogical knowledge should be taught to those who prepare themselves to teach at the elementary level, as well as to teachers who come to us for further studies after having served in schools for a while. In addition to these domains, there is a consensus that elementary mathematics teachers should know something about the domain of children's mathematical thinking. They should be aware of the causes for typical mistakes and should be able to understand children's original ideas about doing mathematics, whether these ideas are correct or incorrect. This is in fact the message of Ball, Hill, and Bass (2005).

I believe that in general, the majority of people who are involved with elementary mathematics teacher training will agree with the above, but when it comes to details, the number of opinions is almost equal to the number of people who are involved in the domain. There are quite obvious reasons for that:

Content Knowledge

The content knowledge depends on the mathematics curriculum which is planned to be covered at the elementary level. Different countries may have different curricula. Even within the same country the curriculum keeps changing over the years. It turns out that mathematics curriculum people have not reached an agreement about questions like: Should we or should we not teach combinatorics or probability at the elementary level? Or, at what grade are we supposed to teach fractions or negative numbers?

During the last 50 years, the elementary mathematics curriculum has been overloaded with some mathematical topics which are beyond the zone of proximal development (Vygotsky, 1978) of the elementary pupils. In addition to that, the dominant approach of the mathematics education community is that mathematical rules and procedures should be explained to the pupils. However, the real mathematical explanations are, sometimes, beyond the pupils' zone of proximal development.

Thus, some mathematics educators came up with alternative explanations which, supposedly, would be understood by the pupils (like explaining multiplication of negative numbers using a "football field" model, see next section). Unfortunately, some of these explanations are quite ridiculous. Their contribution to mathematical education at this level is rather negative.

The following questions still remain to be answered: Should we or should we not eliminate these mathematical topics from the curriculum? Should we or should we not give up the principle that every mathematical rule should be explained? At this point, we can see that content knowledge decisions sometimes depend on pedagogical knowledge available at a given moment. As claimed above, the number of opinions about these issues is almost equal to the number of people who are involved in the domain.

Pedagogical Knowledge

Different mathematics educators of different backgrounds or cultures often have different opinions about the clarity and the efficacy of certain "real-life" models which are supposed to explain why some mathematical operation rules are as they are. The following excerpt (American educator/fall 99) about a "real-life" model to explain multiplying negative numbers can clearly support my claim. It describes an internet chat between schoolchildren:

> Question: Are there any good websites or other resources to help explain negative times negative numbers?
>
> Reactions:
>
> 1. Ds is going through The Key to Algebra, Book 1, and it uses a football field explanation. . . .He does not know much about football and it is confusing.
> 2. Dd said her math book used football as well (Scott Foresman). She knows very little about football and feels his [Ds's] pain.

Children's Mathematical Thinking

There is considerable literature about it (many thousands of pages), which is presented to our teachers and prospective teachers. They do not have the time to study it.

Later on, I will address again the above three knowledge domains. At this point, I would like to ask a question which, in my opinion, is quite important (if not crucial) for the elementary mathematics teachers' work. The answer to this question will determine my answer to the question of what elementary mathematics teachers should know in the above three knowledge domains.

This question is:

To what extent do elementary mathematics teachers have the necessary background to study what we expect them to know so that they will be able to implement it in the tasks that the educational system assigns to them?

The answer will be given in the two following sections.

The van Hiele Hierarchy

At this stage, I would like to introduce to the reader an important work about geometrical thinking which was presented to the community of mathematics educators by two Dutch researchers, Pierre and Dina van Hiele in 1957. The interested reader is referred to Burger and Shaughnessy (1986) for more detailed discussion. I will present the van Hiele hierarchy here very briefly because it is not on the main road of this book and I need it only in order to make a claim about pre-service and in-service teachers. The van Hiele hierarchy has undergone a few transformations. In the beginning it had five levels. Later on, it was reduced to four and in its final form there are only three levels. As I said, the hierarchy relates to geometrical thinking and particularly to dealing with geometrical figures. If, for instance, you point at a square to a pupil who is at the first van Hiele level and ask him: "What is it?" His answer can be: "This is a square." Then you keep asking him: "How do you know?" His answer will be: "It looks like a square."

If you go the same way with a pupil who is at the second level, his answer to the question "How do you know?" will be: "Because it has four equal sides and four right angles." Finally, if you do the same thing with a pupil who is at the third van Hiele level and you ask him: "Is a square also a rectangle?", his answer will be, "Yes, it is, because the definition of a rectangle is a quadrangle that has four right angles and a square does have four right angles." The third level of the van Hiele hierarchy is the theoretical level. As a hierarchy, one cannot be at the third level without having gone through the first two levels.

The van Hiele couple claimed that in many geometry courses the teachers try to teach at the third van Hiele level, whereas the students are only at the first level. This fact is the cause of many failures.

The Location of Student Teachers in the van Hiele Hierarchy

In a study (Gouberman, 2007), based on an adaptation of van Hiele's three geometrical levels to arithmetic, it was found that 63% of the student teachers were below the third level at the beginning of the college arithmetic course. Only 4% of these students improved their location in the three level hierarchy at the end of the arithmetic course. Similar results were obtained in a study by Pandiscio and Knight (2010) which examined the van Hiele level of geometrical understanding of pre-service elementary mathematics teachers, both before and after taking the geometry course required by their teacher preparation program. Results indicate that prior to the course, pre-service teachers do not possess a level of understanding at or above that expected of their target students. The magnitude of the gains (obtained by the end of the course) was not enough to raise the sample population's van Hiele level to that expected of their future K–12 students. There are more studies in which similar results were obtained but I am not mentioning them here because of space restrictions.

I would like to conclude this section about elementary teacher profiles with a quotation from Ball, Hill, and Bass paper (2005): "Many U.S. teachers lack sound mathematical understanding and skills....Mathematical knowledge of most adult Americans is often as weak, and often weaker."

Some Intermediate Recommendations

From what I have said up to now it follows that it is impossible to suggest a uniform list of mathematical topics that prospective teachers should study while preparing themselves to become teachers at the elementary level. However, I would like to present three pedagogical principles that should lead curriculum designers and teacher trainers at the colleges of education:

1. Ausubel's leading principle:
   "If I had to reduce all of educational psychology to just one principle, I would say this: The most important single factor influencing learning is what the learner already knows. Ascertain this and teach him accordingly" (Ausubel, 1968, p. vi). In other words, when instruction is designed, the starting point of the student should determine it. This implies that if the mathematical background of the student is poor we should first improve it and only later move on to more advanced topics.
2. The zone of proximal development principle (Vygotsky, 1978):
   Adapting the zone of proximal development principle to our situation implies that we should not try to teach our students topics which are beyond their intellectual ability. It is worthwhile mentioning that the notion of the zone of proximal development is quite vague. Namely, even if we know "what the learner already knows" we might have difficulties predicting whether the learner is able to cope with a given topic which presumably belongs to the learner's zone of proximal development. For instance, assuming that the learner is familiar with the concept of rational numbers. Will he or she be able to learn the concept of irrational numbers meaningfully?
3. The suitable pace of teaching:
   There is a general tendency to overload syllabi and then, because of the unwritten obligation to cover them, to teach in a pace too fast for the decisive majority of the learners.

Ignoring the above principles leads to meaningless learning. Meaningless learning expresses itself very often in what I have called *pseudo-conceptual* and *pseudo-analytical* behaviors (Vinner, 1997 and Chapter 5 in this book). I will elaborate on it later. At this stage, I would like to claim that for teachers, teaching something that they do not really understand is disastrous. However, this is the case with many elementary mathematics teachers.

Back to the Knowledge Domains Which Elementary Mathematics Teachers Are Supposed to Have

Coming back to the three knowledge domains mentioned above, my recommendations are:

As to mathematical content knowledge, the above three principles can help us determine a list of mathematical topics which can be presented to pre-service and in-service elementary teachers in different social and cultural settings. As explained earlier, these principles cannot lead to a uniform universal curriculum. Giving up a uniform universal curriculum is unacceptable by the majority of influential people involved in national systems of education. Unfortunately, because of the comparative international surveys in science and mathematics, education has become an international competition. Educators and educational policy makers can argue about the advantages and the disadvantages of this fact. However, if as a result of this competition a uniform mathematics curriculum for elementary pupils is suggested, which will also imply a uniform mathematics curriculum for elementary teachers, it will not help mathematical education.

We cannot overcome individual differences through a uniform curriculum. On the contrary, individual differences should be taken care of by differential curricula.

As to pedagogical knowledge, I recommend that "real-life" models and representations should be used only if they are simple and clear. This is true for elementary teachers as well as for elementary pupils.

And as to children's mathematical thinking, since there is no canonical reading list for this topic, I suggest that we present to pre-service and in-service elementary mathematics teachers clear, simple, and straightforward texts rather than sophisticated and complicated studies.

The following three sections will be devoted to answering three additional questions which, in my opinion, are relevant to the elementary mathematics teachers' work:

1. Why do we teach mathematics?
2. What are the views of elementary teachers about the question of what mathematics is?
3. In what ways does the teaching of mathematics serve the ultimate goal of education?

The Common Rhetoric About the Reason for Teaching Mathematics?

In this section, I will answer the first of the three questions I presented above, the question of why we teach mathematics. I consider this question a metacognitive one. Namely, elementary teachers who teach mathematics do not raise it and therefore do not have to answer it. They teach mathematics because it is part of the curriculum which they are supposed to teach. If the question is raised by an external agent, there are ready-made answers to the question, taken from the common rhetoric about it. Usually, curriculum designers present a rationale for teaching the curriculum they recommend. The most beautiful rhetoric for teaching mathematics which I know is the NCTM (1989) rhetoric:

> We live in a mathematical world, whenever we decide on a purchase, choose insurance or health plan, or use a spreadsheet, we rely on mathematical understanding. . . .The level of mathematical thinking and problem solving needed in the workplace has increased dramatically.
>
> . . .Mathematical competence opens doors to productive future. A lack of mathematical competence closes those doors.

This is not the place to elaborate at length about how misleading these claims are. In short, I will say only the following: No doubt mathematical knowledge is crucial to produce and maintain the most important aspects of our present life. This does not imply that the majority of people should know mathematics. Farming is also crucial to at least one aspect of our life—the food aspect, and yet, in developed countries, about 1 or 2% of the population can supply the needs of the entire population. In addition to this argument, if you are not convinced, I recommend to you to look around and to examine the mathematical knowledge of some high-ranking professionals that you know—medical doctors, lawyers, business administrators, and many others, not to mention politicians and mass communication people. Recently, an attempt to refute the above claim that the level of mathematical thinking and problem solving needed in the workplace has increased dramatically came from an unexpected source, a research mathematician, Underwood Dudley (2010) (already mentioned in Chapters 1 and 2) who sampled randomly from the yellow pages eight categories of work places and found no evidence that algebra is required there, "even for training or license."

Another claim in the rhetoric intended to justify the teaching of mathematics is the claim that mathematics is needed for everyday life. However, whenever I ask for specific examples from elementary mathematics teachers who claim it, the only examples they come up with are calculating tips in restaurants, calculating change (this concerns mainly taxi drivers) and cooking (calculating the amount of ingredients for n people, when the amount of ingredients for m people is given in the cook book, where $m \neq n$). There might be other convincing arguments to study mathematics. Underwood Dudley claims that people should study mathematics in order to train their minds. However, there is no experimental evidence which supports the claim that, in non-mathematical domains, people who studied mathematics are better problem solvers than people who did not study mathematics.

Another possible reason for studying mathematics is the application of certain mathematical chapters in sciences (physics, chemistry, biology, etc.). The question here is to what percentage of the population this claim is relevant and whether there are no other ways to reach this percentage rather than impose mathematics on the entire population.

Thus, if the above claims about the need to study mathematics are misleading, why do we teach mathematics and why do our students study mathematics in spite of that? You might suggest that the students believe in these claims although these claims are misleading. I suggest that the students have very good reasons to study mathematics. It is not the necessity of mathematics in their future professional life or in their everyday life. It is because of the selection role mathematics has in all stages of our educational system. Mathematical achievements are required if you want to study in a prestigious place (whether this is a junior-high school, a senior-high school or university). A prestigious school increases your chance to get a good job. Confrey (1995) formulated it quite clearly: "In the vast majority of countries around the world, mathematics acts as a draconian filter to the pursuit of further technical and quantitative studies."

Ultimately, we have a convincing argument to study mathematics. Should we tell it to pre-service and in-service teachers? I believe we should. It is important for a soldier to know the real purpose of a battle in which he or she takes part. He or she should be convinced that there are good reasons for risking his or her life. Intentionally false rhetoric should be morally rejected. I am not claiming that there was a conspiracy to form a false rhetoric about the need to study mathematics. On the contrary, I think that the people who invented this rhetoric really believed in it bona fide. However, beliefs should be re-examined from time to time. The main thing is that teachers will have worthy goals for their endeavor. Is preparing students for crucial examination a worthy goal? I believe it is. Both students and teachers are victims of the same educational reality and as far as we can see, the opportunity to change this reality is very small. For a great part of the younger population, to continue their formal education (generally, not in a domain that requires mathematics) is an important goal. Pupils are expected to progress from the elementary level through the junior-high level to the high-school level and then to college and university. At crucial points of this journey, there are guards who test them on mathematics. If the pupils pass the exams, the guards let them move on. It is a worthy goal to help pupils complete this journey. Of course, there is much more to mathematics. There are intellectual values and educational values. Usually, because of the common way mathematics is taught, pupils are not exposed to them.

What Are the Views of Elementary Teachers About the Question of What Mathematics Is?

The question of what mathematics is, is again a metacognitive question. Generally speaking, people do not seek definitions for the notions they use. The meaning of the decisive majority of concepts in everyday thought is determined by means of examples and not by means of definitions. Some mathematicians, when being asked what mathematics is, prefer to give examples. Among them I can mention Courant and Robbins (1941). Their book is full of mathematical examples. They probably believed that people, who were not mathematics majors but who were eager to know what mathematics was, would be able to understand the mathematical chapters which were presented in the book. Another book which deals with this question is Hersh's book (1997). This is a philosophical book, which presents mathematics and mathematical objects as a human creation, rejecting the platonists' belief that mathematical objects exist independently of the human mind. However, there is no attempt in these books to define mathematics. I believe that (in contrast to other concepts like poetry, art, etc.) it is possible to suggest a definition of mathematics (as I have done in Chapter 8). It is true that in order to understand this definition one should have an appropriate background in mathematics. Maybe a good place to look for simple and general definitions is the dictionary. Webster's Ninth New Collegiate Dictionary suggests that mathematics is "the science of numbers and their operations, interrelations...and of space configurations and their structure. ..." The mathematics that is characterized by this definition includes only arithmetic and geometry. It is not clear whether school algebra is included.

However, elementary school teachers are supposed to know these branches of mathematics.

If you ask them what mathematics is, will they be able to give a similar definition to the one given by the dictionary? Out of 120 elementary mathematics teachers whom I asked, no one gave such an answer. Thus, we can assume that their idea of mathematics is determined by the mathematical experience they had in school and by the mathematics they teach. If you try to generalize this experience, you may suggest that mathematics is a collection of procedures to be used in order to solve some typical questions given in some crucial exams (final course exams, psychometric exams, SAT, etc.).

In order to find out what elementary mathematics teachers really think about mathematics, I made up the following questionnaire and distributed it to my students in a master's degree program for elementary in-service mathematics teachers:

In the teachers' room, during one of the breaks, the literature teacher posed to the mathematics teacher the following question:

"When I was a high-school student I was interested in science. I studied physics, chemistry and biology. I understood that physics described the physical world, chemistry described the chemical world, the elements, and the compounds, and biology described the world of plants and animals. As to mathematics, my impression was that mathematics was a collection of procedures I was supposed to use for homework assignments and for all kinds of mathematics tests, especially the matriculation exams."

A. As a mathematics teacher, what is your opinion about the literature teacher's claim?

    1. I agree.
    2. I partially agree.
    3. I do not agree but I can understand how such an impression is formed.

B. If you do not agree, what would you say to the literature teacher in order to refute her claim?

It turns out that the majority of my students felt that it would be wrong to choose 1 or 2 because of its negative connotation. However, the students who chose 3 gave the following arguments for rejecting 1 and 2:

(I) Mathematics is not only for exams, it is also for real-life situations.
(II) Mathematics teaches you to think.

When asked to specify, in most of the cases they used their right to remain silent.

My question is whether we should tell the elementary teachers what mathematics really is? But before answering this question we should find out whether these teachers have the required mathematical and intellectual backgrounds to understand the answer.

In What Ways Does the Teaching of Mathematics Serve the Ultimate Goal of Education?

Unfortunately, thousands of pages in educational philosophy have been written about the ultimate goal of education. It has also been the theme of hundreds of educational conferences. I say "unfortunately" because we cannot see the forest for the trees. Therefore, I would like to suggest a simple answer to this question. The ultimate goal of education is an educated person.

This claim is, of course, circular. In order to avoid it, I would say that an educated person is a thoughtful person. *Thoughtful* in English is ambiguous. The above Webster dictionary suggests the following: "1: characterized by careful reasoned thinking 2: given to heedful anticipation of the needs and wants of others."

The second part of the above definition of *thoughtful* means, in other words, *considerate*. This can be tied to the *golden rule* mentioned earlier. There are plenty of versions for this rule which come from various cultures and religions. One Jewish version of it, already mentioned earlier, is: "What you hate—do not do to your friend." In order to follow this rule, you should first apply your careful reasoned thinking. Namely, you should carefully analyze everyday situations and determine whether acting in a certain way in these situations will be unpleasant or even harmful to other people. Then, you should control yourself and avoid acting in such a way.

Earlier I mentioned that mathematics, the way it is taught in the majority of schools, focuses mainly on mathematical procedures by means of which typical questions in typical exams can be solved.

Mathematical procedures have negligible importance in everyday life and in the majority of work places. However, as mentioned in Chapter 12, procedures in general, play a crucial role in everyday life and in all work places. Remember that by *procedure* I mean a sequence of actions that should be carried out one after the other. Crossing streets, driving, shopping, turning on dishwashers, dryers, DVD players, etc. are all associated with procedures. This is just an incidental choice out of an infinite list of procedures. Thus, respecting procedures as well as carrying them out precisely and carefully can be recommended as an educational value. As mentioned, not following certain procedures is against the law and may even be dangerous (e.g., crossing an intersection through a red lights while driving a car). Not following other procedures can result in economical damage (e.g., not turning off all the lights and air conditioners when leaving home).

Note that many procedures in everyday life were formed in order to serve the golden rule. For instance, procedures related to standing in line, procedures related to pedestrians and drivers, and procedures related to littering and recycling. Teachers, while teaching mathematical procedures, can refer to procedures in everyday life, and speak about the importance of following these procedures precisely and carefully, the same way as required in mathematics. By doing this, teachers add educational value discussions to their traditional role, which is to cover the syllabus. Within the traditional role, the teacher is a tool of the syllabus. By adding educational discussions to the syllabus, the syllabus becomes an educational tool.

The bottom line of this section brings me back to the typical profile of the elementary mathematics teachers and to the recommendations related to it. While there are serious doubts about the feasibility of presenting to elementary mathematics teachers and prospective teachers some of the mathematical topics mentioned above, there should not be any doubt about asking them to work toward the ultimate goal of education, namely, asking them to be educators.

Unfortunately, in some societies, teachers (and especially elementary teachers) are blamed for not having satisfactory knowledge in science or mathematics. However, if we focus on educational aspects of the teacher's work then the above accusation becomes minor.

Moreover, a counter-accusation should be raised against parents who do not care so much about the education of their children (in the moral sense), but care a lot about their mathematical achievements. They do it not because these achievements are really important for the future life of the children (as adults, 90% of them do not use mathematics, hate mathematics, and do not know mathematics). They do it because mathematical achievements are required for further academic or technical studies of their children.

And as to the elementary teachers—imposing on them mathematical demands which are beyond their mathematical abilities or, if you wish, beyond their zone of proximal development, will bear negative results as explained in the next section.

The Pseudo-conceptual and Pseudo-analytical Behaviors as a Reaction to Exaggerated Intellectual Demands

In Chapter 5, I explained in length what I mean by these two notions (*pseudo-conceptual* and *pseudo-analytical* behaviors). Here, I would like to remind the reader that these behaviors are produced, in some cases, by people who try to show that they know a certain topic, but as a matter of fact they do not know it. In other cases, people really believe that they know but they don't. Here are two anecdotes:

1. The supervision of mathematics education in my country has decided that all elementary mathematics teachers should know certain chapters in probability.

   Thus, teachers who did not have this knowledge were invited to participate in a compulsory course in which some elementary concepts in probability were introduced to them. At the end of the course, among other questions, they were asked to answer the following question: "There are 16 cards in a box. Each card is in an envelope. All the numbers from 1 to 16 appear on the cards (one number per card). Describe an event for which the probability is $\frac{1}{2}$."

   A non-negligent number of the teachers suggested that a right answer to this question was to pull out of the box the card that has 8 on it. When being asked to justify their answer they said: "because $\frac{8}{16}$ is $\frac{1}{2}$." Superficially, it looks like a convincing analytical argument. I classify it as *pseudo-analytical* behavior.

2. An elementary teacher taught how to calculate certain addition exercises by means of Cuisenaire rods. Then, she gave some additional exercises to her class as a homework assignment. One pupil solved all the exercises using his own (mathematically correct) strategy and got the correct answers for all the exercises. However, the teacher marked all his answers as wrong. When the child's mother came to argue with the teacher about her judgment, the teacher's response was that the use of Cuisenaire rods is an essential part of the final result. Here, the teacher did not distinguish between the pedagogical content and the mathematical content. I consider the failure to understand the conceptual difference between the two, a special case of *pseudo-conceptual* understanding.

A Concluding Remark

I would like to conclude this chapter with two comments. The first one is related to the elementary teachers. Observing them in their classes indicates that in most cases they are dedicated people. They do their best to teach mathematics. Sometimes, their best is not good enough mathematically. However, it is useless and pointless to request more than their best. The second comment relates to us, the mathematics education community. My recommendation to focus on the educational aspects of the elementary teacher performance may look to some of us as a threat. Improving mathematical achievements is considered by many of us our ultimate goal. However, I recommend considering other educational values as additional goals of teaching mathematics. There is nothing to worry about. Different educational systems (local, national, and international) will not give up mathematical achievements as a filter for further studying. Hence, improving mathematical achievements will still get the financial support that many of us look for, and mathematics education research will continue to focus on mathematical achievements. My recommendation, therefore, is only to look at things differently and, from time to time, to remind ourselves the real goal of education—an educated adult.

Chapter 14
Proofs by Contradiction or Isn't This an Absurdity?

Quite often when I started a proof by contradiction, at least one student would interrupt me and claim: "How come you ask me to assume that what you said is wrong?" My answer is: "I am not saying it is wrong. I am saying let's assume it is wrong." Still this student would feel quite embarrassed. I can easily understand it. This student has seen some proofs in geometry and developed in his mind an idea how a mathematical proof should look like. All of a sudden there is another kind of proof which looks to him quite weird. However, like many weird things, after a while we get used to them. Here is the place to refer to Fischbein (1987) who spoke about primary intuitions and secondary intuitions. Primary intuitions are results of common sense and everyday experience. For instance, the sun goes around us from east to west. That is our primary intuition. Later on, we learn that our earth is spinning on its axis and that is the reason we have day and night. This is a secondary intuition which takes over the first intuition after a while.

At this stage, we are ready for an additional proof by contradiction (the first one was given in Chapter 11). But in order to prove it, we need a lemma. A lemma is a regular theorem but we call it a lemma because in the course of proving our theorem it has only an auxiliary status.

Lemma: If p is a prime number and a, b are natural numbers and p divides the product $a \cdot b$, then p divides a or p divides b (or both).

At this point, we want to emphasize that this lemma is true only if p is a prime. If p is not a prime and it divides $a \cdot b$, it does not necessarily divide a nor b. For instance, let $p = 6$, $a = 3$, $b = 4$. Then, p divides $3 \cdot 4$ but p does not divide 3 nor 4.

Now back to our lemma. We will not prove it because, as we said, it is not on the main road of proving our theorem; also, its proof is not simple.

Theorem: $\sqrt{2}$ is irrational.

Proof: Let us assume that the negation of the claim is true. That is, $\sqrt{2}$ is rational, and therefore can be written as a simple fraction $\sqrt{2} = \frac{a}{b}$, and that $\frac{a}{b}$ is reduced (if it was not reduced in the first place, we could reduce it). Saying that it is reduced

S. Vinner, *Mathematics, Education, and Other Endangered Species*, Mathematics in Mind, https://doi.org/10.1007/978-3-319-90035-3_14

implies that a and b have no common divisor. Since we assumed that $\sqrt{2} = \frac{a}{b}$, it follows that $2 = \left(\frac{a}{b}\right)^2$. So that:

$$\text{(I)} \quad 2b^2 = a^2.$$

This implies that 2 is a divisor of a^2, namely, of the product $a \cdot a$. 2 is a prime, therefore relying on our lemma, 2 is a divisor of a.

Hence, there is a certain natural number k such that: $a = 2k$. Let us substitute it in (I). We get: $2b^2 = (2k)^2$, or $2b^2 = 4k^2$. Now we reduce it by 2 and we get:

$$\text{(II)} \quad b^2 = 2k^2.$$

This implies that 2 is a divisor of b^2, namely, it is a divisor of the product $b \cdot b$. 2 is a prime; therefore by our lemma 2 is a divisor of b too.

This is a contradiction to the fact that a and b have no common divisor. Our assumptions that the negation of the claim was true lead to this contradiction; therefore, the negation assumption cannot be true. Therefore, the claim must be true.

Q.E.D.

By the way, the above theorem was already known to Pythagoras. This fact embarrassed him a lot because until he found out about it, he believed that the ratio of the lengths of any two given segments is a rational number. But, by his own theorem, the ratio of the diagonal of a square and its side is $\sqrt{2}$ (which, due to the above theorem, turned out to be irrational).

This is another example of primary intuition versus a secondary intuition, as mentioned in Fischbein (1987).

Another theorem that I would like to prove by contradiction is the prime numbers theorem. This theorem was already known to Euclid.

Theorem: There are infinitely many prime numbers.

Proof: Assume that the negation of the claim is true, namely, that there is only a finite number n of prime numbers. Let us denote them by $p_1, p_2, \ldots, p_n$.

Consider now the number $y = p_1p_2\cdots p_n + 1$. y is not divisible by any of the primes $p_1, p_2, \ldots, p_n$ since when dividing it by any of them the remainder is 1. Hence, either y is a prime, or y has a prime devisor that is not among $p_1, p_2, \ldots, p_n$, namely, there is at least one prime that is greater than p_n. This contradicts the assumption that p_n is the last prime. The contradiction is a result of the assumption of the negation of the claim. Therefore, that negation cannot be true, so our claim must be true.

Q.E.D.

When coming back to the subtitle of this chapter "isn't this an absurdity?" I chose it deliberately. It is the English word for *absurdum* in the Latin name of proof by contradiction, *reductio ad absurdum*.

Cognitive Difficulties in Assuming Counter-reality Claims

At the beginning of this chapter, I pointed at the difficulty that some people have to assume counter-reality claims. These difficulties are presented in a humorous way in Ionesco's famous play *The Lesson,* which I already mentioned in Chapter 5. Let us look at the part of the dialogue which is relevant to the above difficulties.

Professor: If you had two noses and I'd plucked one off, how many would you have left?
Pupil: None.
Professor: What do you mean, none?
Pupil: Well, it's just because you haven't plucked one off that I've still got one now. If you had plucked it off, it wouldn't be there anymore.
Professor: . . .Suppose you had only one ear.
Pupil: Yes, and then?
Professor: I stick on another one, how many would you have?
Pupil: Two.
Professor: Good. I stick yet another one on. How many would you have?
Pupil: Three ears.
Professor: I take one of them away, how many ears do you have left?
Pupil: Two.
Professor: Good. I take another one away. How many do you have left?
Pupil: Two.
Professor: No. You have two ears. I take away one. I nibble one off. How many do you have left?
Pupil: Two.
Professor: I nibble one of them off. One of them. . .
Pupil: Two.
Professor: One.
Pupil: Two.
Professor: One!
Pupil: Two!
Professor: One!!
Pupil: Two!!
Professor: One!!
Pupil: Two!!

The pupil can probably not construct in her imagination a picture of her with two noses. Hence, she responds to a situation she can imagine: somebody plucking off her only nose. Unfortunately, she is unable to form the cognitive structures required for the learning situation. The problem is to construct counter-reality "micro-worlds" (herself with two noses). Isn't this a familiar situation in the learning of mathematics? A point without dimensions, an infinite straight line, an infinite plane, infinitely many points between any two given points on a line, a proof by contradiction, where we are supposed to think about a situation the theorem claims cannot happen.

As I pointed out in Chapter 5, at this point the pupil rebels. She no longer agrees to cooperate with the requirement to think about counter-reality examples. Thus, at the end of the play the teacher decides to kill the pupil. In real mathematics classes, if somebody is, metaphorically, supposed to be killed, it is the teacher, not the pupils.

Chapter 15
Visually-moderated Sequences

I saw this notion first in Davis (1984). It is an extremely useful notion to explain procedures in mathematical thinking as well as in everyday situations. The idea is as follows: Assume you are supposed to perform a certain procedure. In mathematics, it can be solving a quadratic equation. Outside mathematics it can be getting from one point to another in your house, on the freeway or at the airport. Usually, we do not remember or do not know all the turning points in our way. However, whenever we reach a turning point the sight of it reminds us what we have to do. Hence, we are visually led by this sight. This is the reason for the name of this concept.

Take for instance the task of solving a quadratic equation. Consider an equation, which has already appeared in this book (in Chapter 12): $\frac{2}{x+1} - \frac{5}{x+2} = 1$.

The moment we see it, we do not simultaneously foresee in our minds all the steps which will bring us to the canonical form of the quadratic equation, namely, to the form $ax^2 + bx + c = 0$. However, we are led by its form to the first step on the way toward its canonical form, which is: Rewrite the fractions with a common denominator. Let us do that. We get:

$$\frac{2(x+2) - 5(x+1)}{(x+1)(x+2)} = 1.$$

At this stage, we are driven by the forms of the numerator and the denominator to remove parentheses in both. We obtain: $\frac{2x+4-5x-5}{x^2+3x+2} = 1$.

We are invited by the form of the last equality to multiply both sides of the equation by the denominator of the left-hand side, and then to combine like terms. We get: $-3x - 1 = x^2 + 3x + 2$. At this stage, we are only one step away from the desirable canonical form of our equation which is: $x^2 + 6x + 3 = 0$.

Now we are invited by the form we got to use the formula which gives us the solutions of the given equation. They are: $x_1 = -3 + \frac{\sqrt{24}}{2}$ and $x_1 = -3 - \frac{\sqrt{24}}{2}$.

S. Vinner, *Mathematics, Education, and Other Endangered Species*,
Mathematics in Mind, https://doi.org/10.1007/978-3-319-90035-3_15

A Case in Which the Visually-moderated Sequence Goes Wrong

Sometimes, unfortunately, if we are not paying enough attention to what we are doing, the visually-moderated sequence can take us to the wrong place. Let us take again the case of the above quadratic equation. When we are taught the theory of the solutions of the quadratic equation, the teacher says: After we reach the canonical form of the quadratic equation, $ax^2 + bx + c = 0$, the solutions are given by the formula $x_1 = \frac{-b+\sqrt{b^2-4ac}}{2a}$ and $x_1 = \frac{-b-\sqrt{b^2-4ac}}{2a}$. Then, the teacher continues:

If $b^2 - 4ac > 0$ then the equation has two solutions, if $b^2 - 4ac = 0$ then the equation has one solution, and if $b^2 - 4ac < 0$ then the equation has no solution.

One student, when solving the above equation in an examination, reached the stage $x^2 + 6x + 3 = 0$ and went on by saying: If $6^2 - 4 \cdot 3 > 0$ then the equation has two solutions, if $6^2 - 4 \cdot 3 = 0$ then the equation has one solution, and if $6^2 - 4 \cdot 3 < 0$ then the equation has no solution.

It is not clear to me whether this student was wrongly driven by a visually-moderated sequence to a context that was relevant to the solution of quadratic equation, but not to the procedure required of him by the examiner, or whether he misunderstood the entire purpose of the entire chapter intended to lead the students to the solution of the quadratic equation.

An Additional Case Where a Visually-moderated Sequence Goes Wrong

Take for instance the algebraic expression $\frac{2}{x+1} - \frac{5}{x+2}$, which appeared above in a quadratic equation. It can appear in another exercise the instruction of which is: Simplify!

Hence, the student starts performing the operations which are dictated to him by the visually-moderated sequence, namely:

$$\frac{2}{x+1} - \frac{5}{x+2} = \frac{2(x+2) - 5(x+1)}{(x+1)(x+2)} = \frac{2x+4-5x-5}{x^2+3x+2} = \frac{-3x-1}{x^2+3x+2}.$$

At this stage, the student is visually moderated to get rid of the denominator, forgetting the original instruction, "Simplify!", and thinking that he or she is solving a quadratic equation. Doing that, they obtain the result: $-3x - 1$ instead of stopping at the previous stage: $\frac{-3x-1}{x^2+3x+2}$. I even saw students going on and writing $x = -\frac{1}{3}$ (the solution of the equation $-3x - 1 = 0$).

Cases in Which the Visually-moderated Sequence Goes Wrong Outside a Mathematical Framework

Assume that you get up in the morning and you move from your bedroom to the kitchen in order to boil water for your first coffee.

Usually, you are led to the kitchen, without even being aware of it, by a visually-moderated sequence. Assume that on your way to the kettle you notice the refrigerator.

This reminds you that there is strawberry ice cream in the freezer and that you are crazy about it (at least I am crazy about it).

Thus, instead of boiling water for your first morning coffee you end up eating strawberry ice cream.

Another case, which is quite typical for old people, is as follows: They decide to go from their bedroom to the study, where their computer is, in order to check their email. However, it is a long way to the study and when they arrive, they forget their original goal. Thus, they stand at the study door asking themselves: "What am I doing here?" In this case, the visually-moderated sequence did not work at all.

Chapter 16
Science Versus Common Sense and Theology

Many of our opinions about the world are formed by our daily experiences. One of them is that the sun makes half a circle in the sky by going from the east in the morning, and approaching the western horizon in the evening. This was the belief of people until Copernicus (1473–1543).

The Copernican Revolution was the paradigm shift from the *Ptolemaic* model of the heavens, which described the cosmos as having Earth stationary at the center of the universe, to the *heliocentric* model with the Sun at the center of the Solar System.

This paradigm contradicted common sense, but moreover, it contradicted the theological belief that Earth is the center of the world. Copernicus was not persecuted by the Church in Poland because the book in which he presented his theory was published immediately after his death in 1543.

However, the book was well known to the scientists communities and the Catholic Church circles. The Italian scientist Galileo (1564–1642) was also involved in the controversy about the heliocentric theory. His telescopic observations supported the heliocentric paradigm, and his writings about it were submitted to the Inquisition. In 1616, the Church Inquisition consultants pronounced heliocentrism heretical and Galileo was ordered not to "hold, defend, or teach in any manner" the Copernican theory regarding the motion of the earth. Galileo obeyed the order for 7 years.

In 1632, Galileo published his book *Dialogue Concerning the Two Chief World Systems*. He was subsequently convicted of heresy and had to spend his remaining years under house arrest.

Fortunately, science has overcome both common sense and theological beliefs. Several centuries later in 1925, a school teacher, John T. Scopes, was accused of violating Tennessee's Butler Act, which had made it unlawful to teach human evolution in any state-funded school. This trial got the name "Scopes' Monkey Trial" because Darwin's evolution theory claimed that we, human beings, are the offspring of monkeys.

Scopes was found guilty and fined $100 ($1366 in 2016), but the verdict was overturned on a technicality.

S. Vinner, *Mathematics, Education, and Other Endangered Species*,
Mathematics in Mind, https://doi.org/10.1007/978-3-319-90035-3_16

Nowadays, in my courses when I mention Darwin's theory I still hear religious students who try to argue that Darwin was wrong. Of course, Darwin's theory contradicts Chapter 1 of Genesis according to which earth animals (including monkeys) and man were created on Friday.

I have a hard time convincing some religious students that Darwin was right (not all religious people reject Darwin.) Usually, these discussions end in a humorous comment: "Perhaps in your family your ancestors were monkeys, not in ours." I smile and add: "It is not an insult to be an animal. Animals are, very often, better than human beings. Wild animals kill only when they are hungry or when they fight for their lives, whereas some people kill just for the sake of killing."

Chapter 17
Rationality

Looking back at what has been claimed by now in Chapters 1–16, it seems to me that what they have recommended is rationality: Thinking rationally and behaving rationally. Now it's your turn, the reader, to ask for a definition. I have expressed my views about the request for definitions for the sake of discussion in Chapter 9 (in the section about potentially negative influences of mathematical thinking). However, in the case of rationality I would like to suggest lexical definitions.

Here are Webster's Ninth New Collegiate Dictionary definitions of rational and related notions:

> **rational** *adj* **1 a:** having reason or understanding **b:** relating to, based on, or agreeable to reason: REASONABLE ⟨a rational explanation, rational behavior⟩
>
> **rationality** *n* **1:** the quality or state of being rational **2:** the quality or state of being agreeable to reason: REASONABLENESS **3:** a rational opinion, belief, or practice—usu. used in pl.
>
> **reason** *n*. . .**2 a** (1)**:** the power of comprehending, inferring, or thinking esp. in orderly rational ways: INTELLIGENCE (2)**:** proper exercise of the mind (3)**:** SANITY **b:** the sum of intellectual powers
>
> **reasonable 1 a:** agreeable to reason ⟨a reasonable theory⟩ **b:** not extreme or excessive ⟨~ requests⟩ **c:** MODERATE, FAIR ⟨a reasonable chance, a reasonable price⟩. . .**2 a:** having the faculty of reason **b:** possessing sound judgment

And from Wikipedia (2006):

> . . .rationality may also denote generally positive attributes, including: reasonable, not foolish, sane and good.

Note that both Webster and Wikipedia agree that rationality and reason, in ordinary language, also have a moral aspect (moderate, fair, and good). Be fair! Be considerate! Be a human being!

As to the aspect of thinking, rational thinking means logically sound, based on clear thought and reason, using systematic methods in reaching a conclusion, and relying on facts.

S. Vinner, *Mathematics, Education, and Other Endangered Species*, Mathematics in Mind, https://doi.org/10.1007/978-3-319-90035-3_17

All these aspects also characterize scientific thinking. So, being rational implies taking science into account.

However, science keeps changing over the course of years. Science in the age of Einstein is essentially different from science in the age of Newton. According to science in the age of Newton, it was considered possible to transform mercury into gold. This was the task of the alchemists. Surprisingly enough, Newton was also an alchemist.

In addition, he was one of the greatest scientists of his period. It is quite possible that some of his great discoveries, such as the force of gravity, were the result of his capability to consider ideas that seemed implausible at first sight. Please note that today everybody is aware of forces that act from a distance. In Newton's days it was not the case, until he discovered gravity.

Nowadays, if a chemist would try to transform mercury or lead into gold he would be kicked out of his laboratory immediately. This is because we have the theory of elements and according to this theory it is impossible to transform elements.

So, as I say, to behave rationally means first of all to take science into account. Science tells us that smoking is unhealthy. It tells us to avoid junk food and it recommends we do physical exercises. But rationality is more than that. Rationality also has, as I claimed above, a moral aspect. Be fair! Be considerate! Be a human being!

Epilogue

The famous American writer Arthur Miller (1915–2005) said the following in a television interview: "Life is long [this was true about him, and about me as well], but art is longer." This means that art will last after we, individuals, die. What Miller claimed about art is also true about other domains, particularly, in our case, about mathematics and mathematics education.

Mathematics and mathematics education were my life's journey. I have invited you, my reader, to join me in that long journey. I hope you did, at least in some of its many sections. I also hope that if you did not like some parts, you got off the train and, perhaps, got on again in a later stage. I hope you are with me now, at the last stop.

I would like to thank you for being with me here and now.

Thanks a lot!

S. Vinner, *Mathematics, Education, and Other Endangered Species*,
Mathematics in Mind, https://doi.org/10.1007/978-3-319-90035-3

References

Ausubel, D. (1968). *Educational psychology: A cognitive view*. New York: Holt, Rinehart, & Winston.

Backhouse, J. K. (1978). Understanding school mathematics: A comment. *Mathematics Teaching, 82*, 39–41.

Ball, D. L., Hill, H. C., & Bass, H. (2005). Knowing mathematics for teaching. *American Educator, 30*(3), 14–17, 20–22, 43–46.

Bauersfeld, H., Krummheuer, G., & Voigt, J. (1998). Interactional theory of learning and teaching mathematics: Related micro-ethnological studies. In H. G. Steiner, & A. Vermondel (Eds.), *Proceedings of the 2nd TME Conference: Foundations and methodology of the Discipline Mathematics Education* (pp. 174–188), Antwerp.

Brousseau, G. (1988). Basic theory and methods in the didactics of mathematics. In P. F. L. Verstappen (Ed.), *Proceedings of the Second Conference on Systematic Co-operation Between Theory and Practice in Mathematics Education* (pp. 109–161). Enschede, The Netherlands.

Burger, W. F., & Shaughnessy, J. M. (1986). Characterizing the van Hiele levels of development in geometry. *Journal for Research in Mathematics Education, 17*(1), 31–48. National Council of Teachers of Mathematics.

Carnap, R. (1932/1959). The elimination of metaphysics through logical analysis of language. In A. J. Ayer (Ed.), *Logical positivism* (pp. 60–81). Glencoe, IL: The Free Press.

Confrey, J. (1995). Student voice in examining "splitting" as an approach to ratio, proportions and fractions. In *Proceedings of the 19th International Conference for the PME* (Vol. 1, pp. 3–29). Recife, Brazil: Universidade Federal de Pernambuco

Courant, R., & Robbins, H. (1941). *What is mathematics?* New York: Oxford University Press.

Davis, P. J., & Hersh, R. (1981). The mathematical experience. In *Birkhäuser Basel*. Boston.

Davis, R. (1984). *Learning mathematics: The cognitive science approach to mathematics education*. London: Croom Helm curriculum Policy Research Series.

Driver, R. (1983). *The pupil as scientist?* Milton Keynes: Open University.

Dudley, U. (2010). What is mathematics for? *Notices of the American Mathematical Society (AMS), 57*(5), 608–613.

Eagleton, T. (2007). *The meaning of life*. Oxford: Oxford University Press.

Erlwanger, S. H. (1973). Benny's conceptions of rules and answers in individually prescribed instruction (IPI) mathematics. *Journal of Children Mathematics Behavior (CMB), 1*(2), 7–26.

Feynman, R. P. (1998). *The meaning of it all*. Cambridge, MA: Perseus Books.

Fischbein, E. (1987). *Intuition in science and mathematics*. Dordrecht, Netherlands: Springer.

Frankl, V. (1959). *Man's search for meaning*. Boston: Beacon Press.

S. Vinner, *Mathematics, Education, and Other Endangered Species*, Mathematics in Mind, https://doi.org/10.1007/978-3-319-90035-3

Gardner, H. (1993). *Multiple intelligences.* New York: Basic Books.
Ginsburg, H. (1977). *Children arithmetic: The learning process.* New York: Van Nostrand.
Goleman, D. (1995). *Emotional intelligence: Why it can matter more than IQ.* New York: Bantam Books.
Gouberman R. (2007). *The relationship between the Developmental Levels of Arithmetic thinking of Pre-service Teachers and the Ability expected from them in Meaningful Learning.* An unpublished Ph.D. thesis in Hebrew, Ben Gurion University of the Negev, Beer Sheva.
Gould, S. J. (1981). *The mismeasure of man.* New York: W. W. Norton and Company.
Hersh, R. (1997). *What is mathematics really.* New York: Oxford University Press.
Ionesco, E. (1951). The lesson. In *Plays*, vol. I, Translated by D. Watson Trans., 1958). John Calder, London, 1958.
Klemke, E. D. (1981). *The meaning of life.* Oxford: Oxford University Press.
Maslow, A. H. (1970). *Motivation and personality* (2nd ed.). New York: Harper & Row.
Matz, M. (1982). Towards a process model for high school algebra errors. In D. Sleeman & J. S. Brown (Eds.), *Intelligent tutoring systems* (pp. 25–50). New York: Academic Press.
Molière. (1670). The would-be gentleman. In *Five plays*, Translated by John Wood. Penguin Classics, 1953. (p. 2 and later).
NCTM. (1989). *Curriculum and evaluation standards for school mathematics.* Virginia: Reston.
Nesher, P., & Teubal, E. (1975). Verbal cues as an interfering factor in verbal problem solving. *Educational Studies in Mathematics, 6*, 41–51.
Pandiscio, E. A., & Knight, K. C. (2010). An Investigation into the van Hiele levels of understanding geometry of preservice mathematics teachers. *Journal of Research in Education, Eastern Educational Research Association, 20*(1), 45–52.
Peressini, A., & Morris, A. (1983). Illinois Universities Test of College Preparatory Mathematics. *Math Tattler News Letter* (The Department of Mathematics, University of Illinois at Urbana-Champaign), pp. 7–9.
Piaget, J., & Inhelder, B. (1958). *The growth of logical thinking from childhood to adolescence.* New York: Basic Books.
Poincare, H. (1914). *Science and method.* London: T. Nelson and Sons.
Pólya, G. (1945). *How to solve it. Doubleday anchor books.* New York: Doubleday & Company.
Popper, K. (1963). *Conjectures and refutations: The growth of scientific knowledge.* New York: Basic Books.
Quine, W. V. O. (1964). *Word and object.* Cambridge: The MIT Press.
Rosnick, P., & Clement, J. (1980). Learning without understanding: The effect of tutoring strategies on Algebra Misconceptions. *The Journal of Mathematical Behavior, 3*(1), 3–27.
Sagan, C. (1995). *The demon-haunted world: Science as a candle in the dark.* New York: Random House, Ballantine Books.
Schoenfeld, A. (1985). *Mathematical problem solving.* New York: Academic Press.
Skemp, R. (1976). Relational understanding and instrumental understanding. *Mathematics Teaching, 77*, 20–26.
Skemp, R. (1977). *The psychology of learning mathematics.* Middlesex: Penguin Books.
Skemp, R. (1979). Goals for learning and quality of understanding. *Mathematics Teaching, 88*, 44–49.
Skemp, R. (1982). Symbolic understanding. *Mathematics Teaching, 99*, 59–61.
Verschaffel, L. (1999). Realistic mathematical modeling and problem solving in the upper elementary school. In J. H. M. Hammers, J. E. H. Van Luit, & B. Csabo (Eds.), *Teaching and learning thinking skills* (pp. 214–240). Lisse: Swets & Zeitlinger.
Vinner, S. (1983). Concept definition, concept image and the notion of function. *The International Journal of Mathematical Education in Science and Technology, 14*(3), 293–305.
Vinner, S. (1997). The pseudo-conceptual and the pseudo-analytical thought processes in mathematics learning. *Educational Studies in Mathematics, 34*, 97–129.
Vygotsky, L. (1986). *Thought and language (English translation).* Cambridge, MA: MIT Press.
Vygotsky, L. S. (1978). *Mind in society.* Cambridge, MA: Harvard University Press.
Watzlawick, P., Weakland, J. H., & Fisch, R. (1974). *Change: Principles of problem formulation and problem resolution.* New York & London: Norton & Company.
Yalom, I. D. (1999). *Momma and the meaning of life.* London: Piatkus.

Index

S. Vinner, *Mathematics, Education, and Other Endangered Species*, Mathematics in Mind, https://doi.org/10.1007/978-3-319-90035-3

FSC
www.fsc.org
MIX
Papier | Fördert
gute Waldnutzung
FSC® C083411

Kolibri 360 GmbH
Ferdinand-Jühlke-Straße 7
99095 Erfurt, Deutschland
produktsicherheit@kolibri360.de